THEORY & APPLICATIONS OF LIAPUNOV'S DIRECT METHOD

WOLFGANG HAHN

English Edition Prepared by
Siegfried H. Lehnigk

Translation by
Hans H. Hosenthien and
Siegfried H. Lehnigk

DOVER PUBLICATIONS, INC.
Mineola, New York

Bibliographical Note

This Dover edition, first published in 2019, is an unabridged republication of the work originally published in the Prentice-Hall International Series in Applied Mathematics by Prentice-Hall, Inc., Englewood Cliffs, New Jersey, in 1963.

Library of Congress Cataloging-in-Publication Data

Names: Hahn, Wolfgang, 1911–1998, author. | Lehnigk, Siegfried H.
Title: Theory and application of Liapunov's direct method / Wolfgang Hahn : English edition prepared by Siegfried H. Lehnigk ; translation by Hans H. Hosenthien and Siegfried H. Lehnigk.
Other titles: Theorie und Anwendung der direkten Methode von Ljapunov. English
Description: Dover edition. | Mineola, New York : Dover Publications, Inc., 2019. | Series: Dover books on mathematics | Text in English, translated from German. | English edition originally published: Englewood Cliffs, N.J. : Prentice-Hall, 1963. | Includes bibliographical references and indexes.
Identifiers: LCCN 2018045874 | ISBN 9780486833606 | ISBN 0486833607
Subjects: LCSH: Lyapunov functions. | Differential equations.
Classification: LCC QA871 .H183 2019 | DDC 515/.39—dc23
LC record available at https://lccn.loc.gov/2018045874

Manufactured in the United States by LSC Communications
83360701   2019
www.doverpublications.com

TABLE OF CONTENTS

1 FUNDAMENTAL CONCEPTS

1. Notations, 1
2. The concept of stability according to Liapunov, 5
3. The idea of the direct method of Liapunov, 11

2 SUFFICIENT CONDITIONS FOR STABILITY OR INSTABILITY OF THE EQUILIBRIUM

4. The main theorems on stability, 14
5. Theorems on instability, 18

3 APPLICATIONS OF THE STABILITY THEOREMS TO CONCRETE PROBLEMS

6. Fundamental remarks on the applications, 22
7. Equations with definite first integrals, 24
8. Construction of a Liapunov function for a linear equation with constant coefficients, 26
9. Simple stability considerations for nonautonomous linear differential equations, 29
10. Equations with linear principal parts, 33
11. Bounds for the initial values, 35
12. Estimates for the stability domain of the parameters, 39
13. The problem of Aizerman and its modifications, 42
14. The problem of Lur'e and its generalizations, 49
15. Estimates for the solutions, 56

TABLE OF CONTENTS

4 THE CONVERSE OF THE MAIN THEOREMS

16. Statement of the problem, 60
17. Uniform stability, 61
18. The inversion of the stability theorems, 68
19. The inversion of the instability theorems, 73
20. On the stability theory of dynamical systems, 75
21. Zubov's method of construction, 78

5 LIAPUNOV FUNCTIONS WITH CERTAIN PROPERTIES OF RATE OF CHANGE

22. Order number and exponential stability, 83
23. Differential equations with homogeneous right hand sides, 88
24. The stability behavior of linear differential equations, 92
25. The order numbers of a linear differential equation, 96

6 THE SENSITIVITY OF THE STABILITY BEHAVIOR TO PERTURBATIONS

26. Stability according to the first approximation, 100
27. The theorem of Liapunov on regular differential equations, 104
28. Total stability, 107

7 THE CRITICAL CASES

29. General remarks on the critical cases, 111
30. The two simplest critical cases, 112
31. Malkin's comparison theorems, 117
32. Special investigations of critical cases, 120
33. The boundary of the stability domain in the parameter space, 123

8 GENERALIZATIONS OF THE CONCEPT OF STABILITY

34. Stability in a finite interval, 126
35. Differential equations with bounded solutions, 129
36. The application of the direct method in general metric spaces, 132
37. Stability in the case of partial differential equations, 137
38. Application of the direct method to differential-difference equations, 139
39. Application of the direct method to difference equations, 146

BIBLIOGRAPHY, 151

AUTHOR INDEX, 175

SUBJECT INDEX, 178

PREFACE TO THE ENGLISH EDITION

The English edition of the "Theorie und Anwendung der direkten Methode von Ljapunov" is, apart from the new Section 35, a translation of the German edition of 1959. I took the opportunity to make a few corrections and, at least, to quote the most recent publications. A closer evaluation of the results, however, was not possible.

Liapunov's method has been recognized in several textbooks, monographs, and introductory presentations, since the German edition of this book was published. In particular, the reader is referred to the following English language publications: Antosiewicz [2], Cesari [1], Cunningham [1], Kalman and Bertram [1], LaSalle and Lefschetz [1], and the summarizing report "AIEE Workshop on Liapunov's Second Method," published by the University of Michigan, Industry Program of the College of Engineering, Ann Arbor, Michigan, 1960.

My thanks are due to Professor Siegfried H. Lehnigk, Huntsville, Alabama, to Mr. Hans H. Hosenthien, Huntsville, Alabama, and to their co-workers, for editing this translation and for valuable comments concerning the text.

W. HAHN

PREFACE TO THE GERMAN EDITION

The fundamental work of A. M. Liapunov (1857–1918) on the stability of motion published in Russian in 1892 and in a French translation in 1907 (Liapunov [1] in the bibliography), originally received only little attention and for a long time was nearly forgotten. Just 25 years ago, these investigations were resumed by some Soviet mathematicians. It was noticed that Liapunov's methods are applicable to concrete problems in Physics and Engineering. Henceforth, as can be seen from the increasing number of publications, mathematicians deal more and more with the stability theory founded by Liapunov. This is particularly true for the so-called *second or direct method* which Liapunov actually used only to establish stability theorems in Theoretical Mechanics. Nowadays, this method is applied to practical problems in the realm of mechanical and electrical oscillations and, particularly, in Control Engineering. On the other hand, it has been recognized that the direct method can serve as a fundamental principle of a general theory of stability comprising considerably more problems than the one ensuing from ordinary differential equations.

The theory of the direct method has been considerably advanced within recent years, and it has reached a certain state of completion. Therefore, it can now be presented in a summarizing progress report. A review offering this subject to a larger circle appears the more indicated since nearly all of the literature was published in Russian, and parts of this literature are difficult to obtain. Among books published outside the USSR, the handbook of Sansone and Conti [1] as well as the textbook of Lefschetz [1] devote only one chapter to Liapunov's method. The excellent textbook of Malkin [19] on the theory of stability is available in German and English translations.

In the subsequent report I have included papers dealing with the direct method which either extend the theory, or use the method as a tool. Hereby, I have endeavored to include the pertinent publications through 1957 as completely as possible. In regard to the fact that the overwhelming majority of these publications is devoted to the derivation and application of stability criteria for ordinary differential equations, I have emphasized the stability theory of ordinary differential equations in the Euclidian phase space, and I have developed it as far as the direct method permits.

PREFACE TO THE GERMAN EDITION

With the exception of occasional references, I have omitted the topological methods (cf., e.g., Elsgolts [1], Nemyckii and Stepanoff [1]), termed "qualitative" methods by the Soviet authors, as well as other methods of stability investigation, such as those developed by Perron [1, 4] and others.

The material is divided in the following manner. The first two chapters contain the elementary part of the theory, the knowledge of which is necessary and practically also sufficient for the applications. A knowledge of the fundamentals of the theory of differential equations and of matrix calculus are the only prerequisites. In these chapters, the primary facts have been fully substantiated; secondary results and extensions have been referred to the "Remarks." Applications in the narrower sense, especially with respect to technical problems, are treated as a whole in Chapter 3. In this manner, I believe, the importance of the problem and of the individual papers is more emphasized than it would be by arranging the results according to strictly systematic points of view. In Chapters 4 to 7, the theory is extended further. Some sections (26, 28, 32, 33) of these chapters, however, are as well of interest for applications.

The concluding Chapter 8 shows that the direct method is not restricted to differential equations, but that with its help an essentially more general stability theory can be established. Chapter 8 requires a more profound knowledge of topology. However, the necessary generalizations are prepared by the formulation of the fundamental definitions and theorems of the previous chapters. In the results, and occasionally also in the proofs, I have mentioned the place of the first publication. Evidently, this cannot be done with full certainty in each case. Particularly important definitions and theorems have been emphasized by putting them in italics. These formulations, however, are not always those of utmost generality.

I have not dealt in large with second-hand presentations. They are included in the references. The recently published book of Zubov [6], however, the only monograph on the direct method published as yet, deserves particular mention. The author, starting from the concept of the dynamical system (cf. Sec. 20) and applying his own method of construction (cf. Sec. 21), arrives at a very elegant derivation of the principal results of the theory. His presentation aims at the utmost generality and, in spite of the title, not at applications in the sense of Chapter 3 of my report.

I would like to express my sincere gratitude to Prof. Dr. F. K. Schmidt, who suggested the writing of this report.

I also offer thanks to Dr. André, Dr. Hornfeck, and Dr. Tietz for their assistance in proof reading as well as for their valuable suggestions and, last, not least, to the publisher for his cooperation and prompt completion of the publication.

W. HAHN

Braunschweig, April 27, 1958

1 FUNDAMENTAL CONCEPTS

1. NOTATIONS

(a) The concepts "stability" and "instability" originated in mechanics as characterizations of the equilibrium of a rigid body. The equilibrium is said to be stable if the body resumes its original position after every sufficiently small displacement. Similarly, a motion is called stable if it is insensitive to small perturbations and to changes in the initial values and in the parameters. Here, in the most simple case, motion means the variation of a point with respect to time, but more generally, we shall understand by a motion the quantities which determine the state of a physical system as a function of time (such as the Lagrange coordinates). A motion can also be interpreted as the trajectory of a point in a space of sufficiently high dimension, thereby permitting the concept of motion to be explained without reference to physical interpretations.

An exact definition of the concept of stability of a motion (which must, of course, comprise the stability of an equilibrium as a special case) can be given in various ways (cf. Moisseev [5]). The definition as formulated by Liapunov [1] has been found to be particularly convenient (cf. Sec. 2). The present work will be based primarily on this definition.

(b) A point of the real, n-dimensional Euclidean space shall be denoted by the coordinates $x_1, \ldots, x_n$. The system of values $\{x_1, \ldots, x_n\}$, denoted by x, is to be treated as a column vector of n components. The row vector which corresponds to x shall be denoted by x^T. Analogously, symbols such as y, z, a, ..., will be used for column vectors with n components. In the case of several vectors $x_1, x_2, \ldots$, the components $x_{1j}, \ldots, x_{nj}$ refer to the vector x_j. Vectors will be denoted by lower case sans serif letters.

The absolute value of the vector x is, as usual, the quantity

$$|\mathsf{x}| = \sqrt{x_1^2 + \cdots + x_n^2} = \sqrt{\mathsf{x}^T\mathsf{x}} \tag{1.1}$$

In addition to the n-dimensional x-space, which is also called *phase space*, we shall refer to the $(n + 1)$-dimensional space of the quantities $x_1, \ldots, x_n, t$, which will be called *motion space*.

(c) The notation $\mathsf{x} = \mathsf{x}(t)$ indicates that the components x_i of x are functions of t. If these functions are continuous, then the point $(\mathsf{x}(t), t)$ of the motion space moves along a segment of a curve as t runs from t_1 to t_2, $t_1 \leq t \leq t_2$. This segment forms a portion of the *motion* of $\mathsf{x}(t)$ in the motion space. The t-axis represents the particular motion belonging to the null vector $\mathsf{x}(t) \equiv 0$.

The projection of a motion upon the phase space is called the *phase curve*, or *trajectory*, of the motion. In this case the quantity t plays the role of a curve parameter. If $\mathsf{x}(t)$ is defined for all $t \geq t_0$, or $t \leq t_0$ there corresponds to this right or left infinite parameter-interval a *branch of the motion*, sometimes called *half-trajectory*.

(d) Scalar functions will be denoted by lower case italic, or, sometimes, Greek letters. Functions of the form $f(x_1, \ldots, x_n)$, or $f(x_1, \ldots, x_n, t)$, which are defined in the phase space or motion space, shall be abbreviated by the notation $f(\mathsf{x})$, or $f(\mathsf{x}, t)$, respectively. Occasionally, several such functions $f_1(\mathsf{x}, t), f_2(\mathsf{x}, t), \ldots$, may be combined to form a column vector $\mathsf{f}(\mathsf{x}, t)$.

Capital sans serif letters denote matrices; If A denotes a matrix, A^T will denote its transpose and A^I its inverse. U will always denote the unit matrix. Capital German letters will be used to denote point sets. The concept "domain" will be used in the general sense, i.e., for open as well as for closed point sets.

(e) The function $f(\mathsf{x}, t)$ is said to belong in $\mathfrak{B}$ to the class $C_0 (f \in C_0)$ if it satisfies a Lipschitz condition at each point of a domain $\mathfrak{B}$, i.e., if in a certain neighborhood of the point (x, t) an inequality of the form

$$|f(\mathsf{x}_1, t) - f(\mathsf{x}_2, t)| \leq m|\mathsf{x}_1 - \mathsf{x}_2|$$

is satisfied, where the constant m is independent of t and x_i. The function f is said to belong to the class C_r if f has continuous partial derivatives with respect to the x_{ij} up to the order r. If there exists for the whole domain $\mathfrak{B}$ a uniform Lipschitz constant m, or if the partial derivatives are uniformly bounded, then $f \in \overline{C}_0$, or $f \in \overline{C}_r$, respectively. C_ω denotes the class of the real analytic functions.

The statement $\varphi(r)$ belongs to the class K means that $\varphi(r)$ is a continuous real function defined in the closed interval $0 \leq r \leq h$, and that $\varphi(r)$ vanishes at $r = 0$ and increases strictly monotonically with r.

(f) The vector differential equation

$$dx/dt = \dot{x} = f(x, t) \tag{1.2}$$

stands for a system of n scalar differential equations of the first order and it might sometimes be equivalent to a single scalar differential equation of the order n. In the following, no distinction will be made in general between vector and scalar differential equations.

If the right hand side of (1.2) is continuous and of such a nature that the existence and uniqueness of the solutions, as well as their continuous dependence on the *initial values* is assured, f shall be said to belong to the class E, $f \in E$. The initial values represent the *initial instant* t_0 (henceforth, we assume $t_0 \geqq 0$) and the *initial point* x_0. If $f \in E$, then $p(t, x_0, t_0)$ will denote that well defined solution which takes on the value x_0 at $t = t_0$, i.e., $p(t_0, x_0, t_0) = x_0$. A constant solution, $p(t, x_0, t_0) \equiv x_0$ is said to be an *equilibrium*, or a *singular point*, of the differential equation. If x_0 is the only constant solution in a neighborhood of x_0, it is called an *isolated* equilibrium.

If the right hand side of (1.2) does not depend on t, or if it is periodic in t, the equation is called *autonomous*, or *periodic*, respectively. Linear differential equations are written in matrix notation as

$$\dot{x} = A(t)x \tag{1.3}$$

(g) Among the scalar functions, the so-called *Liapunov functions* (cf. Definition 4.1) play a particular role. For these functions we shall reserve the letters u, v, and w. For the most part, they are defined in the spherical neighborhood

$$\mathfrak{K}_h : |x| \leqq h \tag{1.4}$$

of the origin of the phase space, or in a half-cylindrical neighborhood

$$\mathfrak{K}_{h,t_0} : |x| \leqq h, \quad t \geqq t_0 \tag{1.5}$$

of the t-axis of the motion space. In all of the following it is assumed that these functions are continuous and that they have continuous partial derivatives with respect to all arguments. The symbols $\mathfrak{K}_h$ and $\mathfrak{K}_{h,t_0}$ shall always have the meaning explained by (1.4) and (1.5). The number t_0 shall mean a fixed, nonnegative and in certain cases sufficiently large constant. To give an exact characterization of a Liapunov function, it is necessary to introduce some definitions.

DEFINITION 1.1: A function $v(x)$ is called *positive (negative) semi-definite* if $v(0) = 0$ and if in a neighborhood $\mathfrak{K}_h$ of the origin $v(x) \geqq 0$ ($\leqq 0$). The case that v vanishes identically is included. If $v(0) = 0$ and $v(x) > 0$ (< 0) for $x \neq 0$, the function is called *positive (negative) definite*.

DEFINITION 1.2: A function $v(x, t)$, defined in a domain $\mathfrak{K}_{h,t_0}$ of the motion space, is called *positive (negative) semi-definite* if $v(0, t) = 0$, ($t \geqq t_0$), and if, with suitable $h_1 \leq h$, $v(x, t) \geq 0$ (≤ 0) in $\Re_{h_1,t_0}$. The function $v(x, t)$ is called *positive (negative) definite* if $v(0, t) = 0$ and if in a domain $\Re_{h_1}(h_1 \leq h)$ of the phase space, a positive definite function $w(x)$ exists, such that the relation

$$v(x, t) \geq w(x) \qquad (\leq -w(x)) \tag{1.6}$$

holds.

Example: $v(x, t) = (1 + t)^{-1}(x_1^2 + x_2^2)$ is positive semi-definite, but not positive definite; however,

$$v(x, t) = (1 + (1 + t)^{-1})(x_1^2 + x_2^2)$$

is positive definite.

DEFINITION 1.3: A function $v(x, t)$, defined in a half-space $t \geq t_0$, is called *radially unbounded* if for each $\alpha > 0$ there is a $\beta > 0$ such that $v(x, t) > \alpha$ whenever $|x| > \beta$ and $t \geq t_0$.

The concept of radial unboundedness has been introduced by Barbašin and Krasovskii [1] with the terminology "v becomes infinitely large."

DEFINITION 1.4: A function $v(x, t)$ is called *decrescent* if the relation

$$\lim v(x, t) = 0 \quad \text{for} \quad |x| \to 0$$

holds uniformly in t. This is equivalent to the existence of a positive definite function $u(x)$ which is independent of t and which satisfies the inequality

$$|v(x, t)| \leq u(x) \qquad (|x| \leq h, t \geq t_0) \tag{1.7}$$

Example: $x_1^2 + tx_2^2$ is not decrescent, but $x_1^2 + x_2^2 \sin t$ is decrescent.

In papers of Liapunov and others who followed him, the terminology "v admits an infinitesimal small upper bound" is used instead of "v is decrescent."

(h) Using the fact that the functions $v(x, t)$ here under consideration are continuous, i.e., that they assume a maximum and a minimum in every closed domain, Definitions 1.1 to 1.4 can be given in a more convenient form.

DEFINITION 1.5: *A function $v(x, t)$ with $v(0, t) = 0$ is called positive (negative) definite if a function $\varphi(r)$ of class K exists such that the relation*

$$v(x, t) \geq \varphi(|x|) \qquad (\leq -\varphi(|x|)) \tag{1.8}$$

is satisfied in $\Re_{h,t_0}$.

DEFINITION 1.6: *A function $v(x, t)$ is called radially unbounded if inequality (1.8) is valid for arbitrarily large h and where $\varphi(r)$ increases unboundedly with r.*

DEFINITION 1.7: *A function $v(x, t)$ is called decrescent, if a function $\psi(r)$ of the class K exists such that in* $\Re_{h,t_0}$

$$|v(x, t)| \leq \psi(|x|) \qquad (1.9)$$

is valid.

Based on these definitions, the following theorems can be stated.

THEOREM 1.1: A function $v(x)$ which vanishes at $x = 0$ and which is continuous there, is decrescent.

THEOREM 1.2: A function $v(x, t)$ which vanishes at $x = 0$, and which has bounded partial derivatives $\partial v/\partial x_i$ in $\mathfrak{K}_{h,t_0}$ is decrescent.

Proof: According to the mean value theorem, we have

$$v(x, t) = x_1 \frac{\partial v(\delta x, t)}{\partial x_1} + \cdots + x_n \frac{\partial v(\delta x, t)}{\partial x_n} \qquad (0 < \delta < 1)$$

Hence, since the partial derivatives are bounded,

$$|v(x, t)| < a(|x_1| + \cdots + |x_n|)$$

THEOREM 1.3: A function $v(x, t)$ is decrescent if there exists for v in a domain $\mathfrak{K}_{h,t_0}$ a power series expansion in the x_i which has no constant term, and which has coefficients uniformly bounded in t.

(i) Let $v(x, t)$ and the differential equation (1.2) be given. If a particular solution $p(t, x_0, t_0)$ of the differential equation (1.2) is substituted for x in $v(x, t)$, then v becomes a function of t alone, along that motion which is determined by this solution. This function of t shall be denoted by $v(t)$. The derivative $\dot{v}$ of $v(t)$ with respect to t is the *total derivative* of $v(x, t)$ with respect to t:

$$\dot{v} = \frac{\partial v}{\partial x_1} f_1(x, t) + \cdots + \frac{\partial v}{\partial x_n} f_n(x, t) + \frac{\partial v}{\partial t} \qquad (1.10)$$

It is called the *total derivative of v for the Eq.* (1.2). Occasionally, it will be more convenient to use the notation $\dot{v}_{(1.2)}$ to refer to the differential equation under consideration. In cases where the total derivative does not exist, the function (1.10) will be replaced by

$$\lim_{\Delta t \to 0+} \sup \frac{v(x(t + \Delta t), t + \Delta t) - v(x(t), t)}{\Delta t} \qquad (1.11)$$

2. THE CONCEPT OF STABILITY ACCORDING TO LIAPUNOV

Let the components $y_i(t)$ of a motion $y(t)$ continuously depend on certain parameters $a_1, a_2, \ldots$. A particular constellation of such parameters shall be denoted by a. It defines a point in the *parameter space*. Let a_0 be a fixed point in the parameter space and a a variable point belonging to a neighborhood $\mathfrak{U}$ of a_0. $y(t, a_0)$ and $y(t, a)$ shall be the corresponding motions. The motion of $y(t, a_0)$ is called the *unperturbed motion;* all other motions $y(t, a)$ are called *perturbed motions*. According to Liapunov, we introduce:

DEFINITION 2.1:

(a) *The unperturbed motion* $y(t, a_0)$ *is said to be stable (with respect to the parameter set* $\mathfrak{U}$ *and the initial instant* t_0*) if for each* $\epsilon > 0$ *there is a* $\delta > 0$ *such that the inequality*

$$|y(t_0, a_0) - y(t_0, a)| < \delta \tag{2.1}$$

implies

$$|y(t, a_0) - y(t, a)| < \epsilon \qquad (t \geqq t_0) \tag{2.2}$$

The number δ depends on ϵ; however, in general it will also depend on t_0.

(b) *The unperturbed motion is said to be quasi-asymptotically stable if the relation*

$$\lim (y(t, a_0) - y(t, a)) = 0 \qquad (t \to \infty) \tag{2.3}$$

holds for all a *of a certain neighborhood* $\mathfrak{U}_0 \subseteq \mathfrak{U}$ *of* a_0. A motion which is stable, but not asymptotically stable is said to be *weakly stable*.

The concept of quasi-asymptotic stability has been introduced by Antosiewicz [2].

DEFINITION 2.2: *The unperturbed motion is called asymptotically stable if it is both stable and quasi-asymptotically stable.*

Both properties combined in this definition are independent of each other (cf. the example given below).

DEFINITION 2.3: *The unperturbed motion is called unstable if it is not stable.* In this case there exists a fixed $\epsilon > 0$ with the following property: there exists a sequence of numbers $t_1, t_2, \ldots \to \infty$ and a sequence of vectors $y(t, a_n)$ such that

$$|y(t_n, a_0) - y(t_n, a_n)| \geqq \epsilon$$

although

$$\lim_{n \to \infty} \{y(t_0, a_0) - y(t_0, a_n)\} = 0$$

Examples: Each motion of the family $y = a \sin t$ $(-\infty < a < \infty)$ is weakly stable if $t_0 \neq k\pi$ (k integer). For $t_0 = k\pi$ the motions are unstable. Each motion of the family $y = ae^{-t} \sin t$ is asymptotically stable for $t_0 \neq k\pi$. For $t_0 = k\pi$ it is quasi-asymptotically stable, but not stable.

Each motion of the family $y = \sin \omega t$ $(-\infty < \omega < \infty)$ is bounded, but unstable.

The motions of the family $y = e^{-\alpha t} \cos \omega t$ belonging to the parameter interval $-\infty < \omega < \infty$, are quasi-asymptotically stable if $\alpha > 0$. Each motion is unstable if $\alpha \leqq 0$.

In the study of actual motions, the disjunction of stability and instability is of less importance than the disjunction of asymptotic stability and instability. Weak stability is a relatively rare special case. Most families contain asymptotically stable and unstable motions but no weakly stable motions. The theory of stability is primarily concerned with motions determined by ordinary differential equations, where the initial values serve as parameters. When speaking of stability of these motions, the appendage "with respect to the initial values" will be omitted. Let

$$\dot{y} = f(y, t) \tag{2.4}$$

be the differential equation under consideration, and let v be a particular solution determining the unperturbed motion. If we put

$$x = y - v \tag{2.5}$$

we obtain the differential equation

$$\dot{x} = f(x + v, t) - f(v, t) \tag{2.6}$$

which is called the *differential equation of the perturbed motion*. This equation has the trivial solution $x = 0$ as an equilibrium (Sec. 1 (f)). However, Eq. (2.6) is often more complicated than the original equation (2.4).

The majority of the theorems of stability theory provide statements about the stability or instability of the equilibrium. Therefore, in the sequel we shall make the assumption that the differential equations, considered as equations of a perturbed motion, shall have the trivial solution as an *isolated equilibrium*. In the following, this isolated equilibrium will be called simply *the* equilibrium. Unless otherwise stated, we shall furthermore assume that the right hand sides of the differential equations belong to the class E (cf. Sec. 1 (f)) in a certain domain $\mathfrak{R}_{h,t_0}$. Occasionally, differential equations whose right hand sides do not satisfy these conditions (which are, for instance, not one-valued in a neighborhood of the origin) can be made amenable to the theory by introducing new variables (cf. Dubošin [1]). If possible, we shall denote the dependent variable by x, and write

$$\dot{x} = f(x, t) \qquad (f(0, t) = 0, f \in E) \tag{2.7}$$

Definitions 2.1 and 2.2 will now be given in a form of particular interest for applications. It is convenient to define separately the two properties resulting in the concept of asymptotic stability.

DEFINITION 2.4:
(a) *The equilibrium of the differential equation (2.7) is said to be stable if there exists for each $\epsilon > 0$ a number $\delta > 0$ such that the inequality*

$$|x_0| < \delta \tag{2.8}$$

implies

$$|p(t, x_0, t_0)| < \epsilon \qquad (t \geq t_0) \tag{2.9}$$

(b) *The equilibrium is said to be quasi-asymptotically stable if there is a number $\delta_0 > 0$ such that from $|x_0| < \delta_0$ the relation*

$$\lim p(t, x_0, t_0) = 0 \qquad (t \to \infty)$$

follows.

In this case, there exists for each $\eta > 0$ a number $\tau = \tau(\eta)$ such that $|x_0| < \delta_0$ implies the inequality

$$|\mathsf{p}(t, x_0, t_0)| < \eta \qquad (t > t_0 + \tau) \tag{2.10}$$

The number τ depends in general on t_0 and x_0 (cf. Sec. 17).

(c) *The equilibrium is said to be asymptotically stable if it is both stable and quasi-asymptotically stable.*

Definition 2.4 permits the following interpretation. Inequality (2.9) defines a half-cylindrical neighborhood $\mathfrak{K}_{\epsilon,t_0}$ of the t-axis of the motion space and inequality (2.8) defines an n-dimensional spherical domain $\mathfrak{K}_\delta$ in the hyper-plane $t = t_0$ with the center at $(0, t_0)$. The condition for the stability of the equilibrium means that a motion can be forced to remain in the domain $\mathfrak{K}_{\epsilon,t_0}$ by guiding the motion for $t = t_0$ through the spherical domain $\mathfrak{K}_\delta$. In the case of asymptotic stability, all motions originating in the spherical domain remain in a neighborhood of the t-axis contracting towards the t-axis.

The properties explained in Definitions 2.4(a) and 2.4(b) are independent of each other, as can be seen from examples given by Massera [4].

In the case of asymptotic stability, the set of all points (x_0, t_0) from which motions originate, satisfying the relation (2.10), forms the *domain of attraction* of the equilibrium. The dependence on t_0 is generally of no importance (cf. the following Remark 1). Therefore, in general, we shall understand by the domain of attraction only that part of the phase space in which decaying motions originate. The domain of attraction can also be characterized by the fact that *every* motion originating in it is asymptotically stable.

If relation (2.10) is valid for *all* points x_0 from which motions originate, we shall say that the equilibrium is *asymptotically stable in the large* (Aizerman [4], Krasovskii [7]). If relation (2.10) holds for all points of the phase space, the equilibrium is said to be *asymptotically stable in the whole*. (Barbašin and Krasovskii [1, 2]). LaSalle [2] proposed "complete stability." The distinction between asymptotic stability in the large and asymptotic stability in the whole has often been obliterated by inaccurate translations of the Russian terminology. However, it becomes important in cases where Eq. (2.7) is not defined for all points of the phase space. Refinements of the concept of stability which are necessary for further extensions of the theory (uniform stability, etc.) will be introduced later (Secs. 17 and 22).

With respect to the equilibrium, Definition 2.3 can be expressed in the following form.

DEFINITION 2.5: *The equilibrium is called unstable if there exists a number $\epsilon > 0$ with the following property: There exist a sequence of numbers*

$\tau_1, \tau_2, \ldots \to \infty$ and a null sequence of initial points $x_1, x_2, \ldots \to 0$ such that

$$|\mathsf{p}(t_0 + \tau_n, x_n, t_0)| \geqq \epsilon \qquad (n = 1, 2, \ldots)$$

A stronger form of instability is described by the following property: There exist a number $\epsilon > 0$, a null sequence $x_n \to 0$, and for every n an unboundedly increasing sequence $\tau_{n,m}$ such that

$$|\mathsf{p}(t_0 + \tau_{n,m}, x_n, t_0)| \geqq \epsilon \qquad (m, n = 1, 2, \ldots)$$

The equilibrium is unstable, for instance, if solutions which increase unboundedly with t originate in every neighborhood of the origin. But this is not at all a necessary condition.

A special case of instability occurs when every motion tends away from the equilibrium. This case is described in:

DEFINITION 2.6: The equilibrium is said to be *completely unstable* if there exists a number $\epsilon > 0$ with the following property: After finite time, *each* motion $\mathsf{p}(t, x_0, t_1)$ reaches the sphere $|x| = \epsilon$, where $0 < |x_0| < \epsilon$, and $t_1 \geqq t_0$.

Remarks:

1. The initial instant t_0 appears as a parameter in the previous definitions, and the number δ in (2.8) generally depends on t_0. If for t_0 a number $\delta = \delta(\epsilon, t_0) > 0$ can be found which corresponds to the stability requirement, then there also exists a positive $\delta(\epsilon, t_1)$ for each $t_1 > t_0$. For, by the assumption about the right hand side of (2.7), each motion can be traced back from t_1 to t_0 (cf. Kamke [1], Sec. 16). The motions originating from the spherical domain $|x_0| < \delta(\epsilon, t_0)$ intersect the hyperplane $t = t_1$ in a domain $\mathfrak{B}$ which in general is not a sphere about $x = 0$. However, the point $x = 0$ is certainly an interior point of $\mathfrak{B}$ and $\mathfrak{B}$ certainly contains a spherical domain $|x| < \delta(\epsilon, t_1)$ with nonvanishing radius.

2. A slight modification of the definitions becomes necessary if the assumptions about the right hand side of (2.7) are weakened so that only the existence, but not the uniqueness of the solutions is guaranteed. "Motion" is then understood to mean the union of all integral curves originating from the point (x_0, t_0). The "distance of the motion from the origin" has to be defined by sup $|x(t)|$, where $x(t)$, for fixed t, passes through all points of the solutions $\mathsf{p}(t, x_0, t_0)$ of which the motion is composed. (Kurzweil [3]; cf. also Sec. 17, Remark 5 and Sec. 35.)

3. The quantities occuring in stability problems are in general real. However, the definitions remain valid for equations of the type (2.4) and (2.7) in the complex domain, provided the independent variable t is real (Vejvoda [1]). The complex case can always be reduced to the real case by separation of the real and imaginary parts.

4. Moisseev [2, 4] defined the concept of "probability of stability" to more precisely describe the situation in the case of an unstable equilibrium. By an "S-motion," Moisseev understands a solution of (2.7) such that $|x_0| < \delta$ implies the inequality

$$|\mathsf{p}(t, \mathsf{x}_0, t_0)| < \varphi(\delta)$$

for all $t \geq t_0$, where $\varphi(r)$ is any function of the class K (Sec. 1 (e)). Let $M_S(\delta)$ be the measure of the points of the spherical domain $|\mathsf{x}| < \delta$ from which S-motions originate and let $S(r)$ be the measure of the sphere $|\mathsf{x}| = r$. The *probability of stability* is defined by the limit value of the quotient $M_S(\delta)/S(\delta)$ for $\delta \to 0$. This limit value depends on the choice of the function $\varphi(\delta)$. Stepanoff [1] defined a similar concept for autonomous motions only: Let $E(R, r)$ be the measure of the set of those points x_0 of the spherical domain $|\mathsf{x}| < r$ for which $\sup |\mathsf{p}(t, \mathsf{x}_0, 0)|$ for $t \geq 0$ is not greater than R. The probability of stability is defined as

$$\lim_{R \to 0} \lim_{r \to 0} \frac{E(R, r)}{S(r)} \qquad (2.11)$$

provided this limit exists.

Obviously, the probability of stability as defined by (2.11) is equal to 1 for the case of stability in the sense of Liapunov. However, the converse is not valid. For example, the trajectories of the system of differential equations

$$\begin{aligned}\dot{x} &= x^2 - y^2 \\ \dot{y} &= 2xy\end{aligned} \qquad (2.12)$$

are the circles passing through the origin and with centers located on the y-axis, the origin itself, and the entire x-axis. Consequently, the equilibrium is unstable. The measure $E(R, r)$ is equal to the area of the two lunae which are cut from the circle with radius r about the origin by the circles about $(0, \pm R/2)$. Independent of R, we have here

$$\lim_{r \to 0} \frac{E(R, r)}{S(r)} = 1$$

5. If the left hand sides of the inequalities (2.8) and (2.9) are replaced by

$$\left\{\sum_{i=1}^{m} x_i^2(t_0)\right\}^{1/2} \quad \text{or} \quad \left\{\sum_{i=1}^{m} x_i^2(t)\right\}^{1/2}$$

respectively, where $m < n$, i.e., if not all of the variables are considered, one arrives at the concept of *stability with respect to a subset of the variables*. This concept has been more closely investigated by Rumiancev [6]. (Cf. also Zubov [10].)

6. Kats and Krasovskii [1] introduced the concept "stable with probability." This concept is of importance in connection with differential equations whose right hand sides depend on a statistically given function. In such a case, inequality (2.8) has to be replaced by the following statement: The probability for (2.8) being satisfied can be brought as close to unity as desired by suitable choice of δ. In the case of asymptotic stability with probability, the probability for (2.8) being satisfied approaches unity.

3. THE IDEA OF THE DIRECT METHOD OF LIAPUNOV

The direct, or second method of Liapunov attempts to make statements on the stability of the equilibrium without any knowledge of the solutions of the differential equations, i.e., without using the explicit form of the perturbed motions. The name "second method" is of historical origin. Liapunov [1] also used a "first method." The first method comprises all procedures in which the explicit form of the solutions is used, especially when represented by infinite series. In this report, we will not refer to the first method. The second method attempts to make stability statements directly by using (in addition to the differential equations) suitable functions which are defined in the phase space, or in the motion space. These functions are usually called *Liapunov Functions*. In general, the sign of the Liapunov function, and the sign of its time derivative for the differential equation have to be considered. The method admits a geometrical interpretation which very likely was originally given by Chetaev [8] and which appears to be particularly useful in applications. Subsequently, the method shall be illustrated in the most simple case. The following considerations do not replace a proof. Let $x(t)$ be the general solution of two autonomous differential equations of first order. We construct the totality of all phase curves which schlichtly cover the phase plane, or at least a neighborhood of the origin. Furthermore, let $v(x)$ be a positive definite function such that $v(x) = c = $ const defines for sufficiently small positive constants c a family of closed curves which also schlichtly cover a neighborhood of the origin (cf. the following Remark 2). The origin itself is located in the interior of each curve and corresponds to the value $c = 0$. We shall assume that all phase curves originating from points of the circular disc $|x| = r$ cross the curves $v(x) = c$ from the exterior towards the interior if we go along these curves in the direction of increasing values of the parameter t. Then we can conclude that these phase curves approach the origin arbitrarily closely as t increases, i.e., the equilibrium is asymptotically stable. Analytically, the behavior of the phase curves can be described in the following manner: The function $v(t) = v(x(t))$ decreases monotonically as t increases. Its total derivative $\dot{v}$ for the differential equation must always be negative for sufficiently small values of $|x|$. For asymptotic stability in the whole, we must demand $\dot{v} < 0$ for all x and, in addition, that the curves $v(x) = c$ remain closed for arbitrarily large c. For example, this does not hold for

$$v = x_1^2 + (\exp(-x_2^2) - 1)^2$$

If $\dot{v} < 0$ holds only for $|x| > r > 0$, then asymptotic stability cannot be guaranteed, but, the solutions are at least bounded.

If the function $v(\mathsf{x})$ is indefinite, there exist in each neighborhood of the origin points where $v < 0$. The domain $v < 0$ is bounded by the curve $v = 0$. Assume the case that $\dot{v} < 0$ along a phase curve originating somewhere on the boundary curve $v = 0$. Then this curve penetrates the domain $v < 0$ as t increases, and it can never again reach a point where $v = 0$; the equilibrium is unstable. Apparently, less information is required for this conclusion than is necessary in the case of stability. The inequality $\dot{v} < 0$ needs to hold only in the domain $v \leqq 0$. These considerations can be easily extended to systems of higher order as well as to nonautonomous equations. In the latter case, the motion space has to be considered. For the rigorous foundation of the method (cf. Sec. 4 ff), the geometrical interpretation is not sufficient.

Remarks:

1. The main characteristic of the direct method is the introduction of the function v, by which a generalized distance from the origin of the phase space, or from the t-axis of the motion space, is defined. The method is characterized by the *systematic application of such functions for the deduction of general statements*, particularly about the stability behavior of the trivial solution (equilibrium). Only recently, Reissig, Yoshizawa, and others used the method to derive *statements about the boundedness* (cf. Sec. 35). "Liapunov functions" have been applied in several specific investigations; by Bendixon [1], for example, to study the differential equation $y' = f(x, y)$ in the neighborhood of a singular point, by Levinson [1], and by Reuter [1, 2] to prove the boundedness of the solution of certain equations of oscillations. But in these specific investigations no systematic use was made of Liapunov's theory.

2. Only a certain part of the curve, or surface, $v = c$ was used in the previous geometrical considerations; it can be characterized as follows: We connect the points of a sphere $|\mathsf{x}| = r$ of sufficiently large radius r with the origin by all possible continuous paths, and we consider the locus of all those points for which the paths first intersect the surface $v = c$ when traversing the path from the origin. In the sequel, the surface $v = c$ shall always be understood as the so-defined *cycle* (Chetaev [8]). If v depends explicitly on t, we define the cycle $v = c$ in the following manner: We consider for fixed $t \geqq t_0$ the cycle $w = c$, where w is the function defined in (1.6), and we construct the locus of all points at which the paths drawn from the origin to the cycle $w = c$ intersect the surface $v = c$ for the first time. Apparently, the cycle $w = c$ "comprises" the cycle $v = c$.

3. As already mentioned in the autonomous case, the function $v(\mathsf{x})$ can be considered as a generalized distance from the origin of the phase space. If a function $u(\mathsf{x})$ is written in the form

$$u = f(\mathsf{x})/g(\mathsf{x})$$

it can be interpreted as the angle-coordinate in a suitably chosen cylindrical coordinate system under certain assumptions about the pencil of surfaces determined by $f(\mathsf{x}) - cg(\mathsf{x}) = 0$, i.e., by $u = c = \text{const}$. The simplest example for $n = 3$ is $u = x_1/x_2$; the pencil of surfaces consists of planes. The condition $\dot{u} > 0$ means that the corresponding motion rotates about the axis of the pencil of surfaces always in the same rotational sense. If the motions always remain within a torus-like domain, then it is possible to infer the existence of almost periodic motions which characterize a quasi-oscillatory state. The precise analytical formulation of this situation was given by Nemyckii [3, 4] who named u a *rotating Liapunov function*. These concepts are not needed in the phase plane, since the existence of periodic solutions is already guaranteed if the phase curves remain in an annular domain.

4. By means of some examples, Reissig [2, 7] showed that it is sometimes possible to use discontinuous Liapunov functions. For example, let

$$v = a_1 x^2 + y^2 \quad (xy < 0), \qquad v = a_2 x^2 + y^2 \quad (xy > 0)$$

If $a_1 \neq a_2$, the consideration of the first part of this section cannot be carried out, since the curves $v = $ const are not closed, but have jumps at the x-axis. However, if we know that the decrease of $|v|$ along a phase curve in the upper, or lower half-plane, respectively, is greater than the jump of v while crossing the x-axis, then the above explained conclusion does hold. But this procedure can only be applied if certain properties of the phase curves are already known.

2 SUFFICIENT CONDITIONS FOR STABILITY OR INSTABILITY OF THE EQUILIBRIUM

4. THE MAIN THEOREMS ON STABILITY

The following theorems concern the equilibrium of the differential equation (2.7),

$$\dot{x} = f(x, t) \qquad (f \in E)$$

and, together with the theorems of Sec. 5, form the actual core of the direct method. For the sake of simplicity, each statement in which a Liapunov function v occurs will be stated only for one sign of that function. With each statement based on v there can be associated a completely equivalent statement based on $-v$.

THEOREM 4.1 (Liapunov [1]): *The equilibrium is stable if there exists a positive definite function $v(x, t)$ such that its total derivative $\dot{v}$ for the differential equation (2.7) is not positive.*

Proof: By assumption, a function $\varphi(r)$ of class K exists such that in a domain $\mathfrak{R}_{h,t_0}$

$$v(x, t) \geq \varphi(|x|) \qquad (4.1)$$

Now, let $\epsilon < h$ be given, and let x_0 be so chosen that the inequalities

$$|x_0| < \epsilon \quad \text{and} \quad v(x_0, t_0) < \varphi(\epsilon) \qquad (4.2)$$

are satisfied simultaneously. Since v is continuous and equal to 0 at the origin, such a vector x_0 always exists. Then it certainly follows that $|p(t, x_0, t_0)| < \epsilon$ for small values of $t - t_0$. If there were a $t_1 > t_0$ for which $|p(t_1, x_0, t_0)| = \epsilon$, it would follow that

$$v(t_1) = v(p(t_1, x_0, t_0), t_1) \geq \varphi(|p(t_1, x_0, t_0)|) = \varphi(\epsilon) \qquad (4.3)$$

in contradiction to the inequality
$$v(t_1) \leqq v(t_0) < \varphi(\epsilon) \qquad (t_1 > t_0)$$
which follows from $\dot{v} \leqq 0$.

THEOREM 4.2 (Liapunov [1]): *The equilibrium is asymptotically stable if there exists a positive definite, decrescent function $v(\mathbf{x}, t)$ such that its total derivative for (2.7) is negative definite.*

Proof: The equilibrium is at least stable, since the assumptions of the theorem include those of Theorem 4.1. Consequently, it is possible to choose a number δ so that the motions determined by $|\mathbf{x}_0| < \delta$ remain in the interior of the domain $\mathfrak{K}_{h_1,t_0}$ for a given $h_1 \leqq h$. According to the assumption, there also exist two functions $\chi(r)$ and $\psi(r)$ of the class K such that for $t \geqq t_0$

$$\dot{v} \leqq -\chi(|\mathsf{p}(t, \mathsf{x}_0, t_0)|), \qquad v \leqq \psi(|\mathsf{p}(t, \mathsf{x}_0, t_0)|) \tag{4.4}$$

$v(t)$ decreases monotonically as t increases, and, since $v(t) \geqq 0$,

$$\lim_{t \to \infty} v(t) = \lim_{t \to \infty} v(\mathsf{p}(t, \mathsf{x}_0, t), t) = v_0 \geqq 0$$

exists. If v_0 were positive, then $\psi(|\mathsf{p}|)$ would be positive for all $\mathsf{p}(t, \mathsf{x}_0, t_0)$ with sufficiently small $|\mathsf{x}_0|$, since $\psi(|\mathsf{p}|) \geqq v_0$. Hence, we would have

$$|\mathsf{p}(t, \mathsf{x}_0, t_0)| \geqq p_0 > 0$$

Consequently,

$$\dot{v}(t) = \dot{v}(\mathsf{p}(t, \mathsf{x}_0, t_0), t) \leqq -\chi(p_0) \qquad (t \geqq t_0)$$

and by integration

$$v(t) = v(t_0) + \int_{t_0}^{t} \dot{v} \, dt \leqq v(\mathsf{x}_0, t_0) - (t - t_0)\chi(p_0) \tag{4.5}$$

In the case $p_0 > 0$, inequality (4.5) would lead to a contradiction, since $v(t) \geqq 0$. Therefore, $p_0 = 0$, and $v_0 = 0$, i.e., $\lim \mathsf{p}(t, \mathsf{x}_0, t_0) = 0$.

THEOREM 4.3 (Barbašin and Krasovskii [2]): *The equilibrium is asymptotically stable in the whole if there exists a function $v(\mathbf{x}, t)$ which is everywhere positive definite, radially unbounded, and decrescent, and whose total derivative for (2.7) is negative definite.*

The proof is similar to the proofs of Theorems 4.1 and 4.2 with h now permitted to be arbitrarily large.

DEFINITION 4.1: *A function v which satisfies the conditions of the theorems of Secs. 4 and 5 is called a Liapunov function for the differential equation (2.7).*

Remarks:
 1. According to Moisseev [1], one can dispense with Liapunov's original requirement on the continuity of the function $\dot{v}$ and admit finite jumps.
 2. It suffices to require only the continuity of the right hand side of differential equation (2.7), since the uniqueness of the non-zero solutions is

not used in the proofs. However, the concept of motion has to be defined in a correspondingly general sense (Sec. 2, Remark 2). The equilibrium must naturally be the only existing one in $\Re_{h,t_0}$.

3. If a positive definite, decrescent function $v(\mathsf{x}, t)$ and a negative definite function $v_1(\mathsf{x}, t)$ can be found such that, with increasing t, the function $u = \dot{v} - v_1$ tends uniformly towards zero in every fixed domain $0 < h_1 \leq |\mathsf{x}| \leq h_2 < h$, then the equilibrium is stable (Malkin [5]) and it is even asymptotically stable (Massera [1]). The existence of a positive definite Liapunov function with negative definite derivative without the additional condition that this function is decrescent does not even guarantee the quasi-asymptotic stability of the equilibrium. Thus, the only possible conclusion is that $\mathsf{p}(t, \mathsf{x}_0, t_0)$ approaches zero on a suitably chosen sequence $t = t_n (t_n \to \infty)$ (Antosiewicz [2], Massera [4], cf. also Sec. 19).

Another way of weakening the condition "$\dot{v}$ negative definite" was given by Kuzmin [1]. His theorem corresponds to the instability theorem of Chetaev [2, 6] stated in Sec. 5 after Theorem 5.3.

4. As inequality (4.5) shows, rather than requiring $\dot{v}$ to be definite, it is sufficient that an inequality of the form

$$\dot{v} \leq -\xi(t)\chi(|\mathsf{p}|)$$

holds for sufficiently small $|\mathsf{p}|$, where $\xi(t) \geq 0$ and where the integral

$$\int_{t_0}^{t} \xi(\tau)\, d\tau$$

increases beyond all bounds.

5. Let $\varphi(r, t)$ be continuous for $0 \leq r \leq h$, $t \geq t_0$, and be of such a nature that the scalar differential equation

$$\dot{y} = \varphi(y, t)$$

has unique solutions in the domain just mentioned (one of these solutions is the trivial one). Let $v(\mathsf{x}, t)$ be positive definite and let

$$\dot{v} \leq \varphi(v(\mathsf{x}, t), t) \qquad (4.6)$$

If the equilibrium $y = 0$ is stable, or asymptotically stable, the corresponding statement holds also for the equilibrium of (2.7) (Corduneanu [3]). A special case of this theorem (Massera [4]) is: Let $v(\mathsf{x}, t)$ be positive definite, let $\gamma(r)$ be a function of the class K and let $\dot{v} \leq \gamma(v(\mathsf{x}, t))$: then the equilibrium is asymptotically stable.

6. In the assumptions of Theorem 4.2, the condition "decrescent" can be replaced by "$\mathsf{f}(\mathsf{x}, t)$ bounded" (Marachkov [1], Massera [1]). However, one cannot dispense with both conditions simultaneously (Massera [1]).

7. Let the equation be autonomous, $\dot{\mathsf{x}} = \mathsf{f}(\mathsf{x})$. The existence of a positive definite function $v(\mathsf{x})$ with the following properties is then sufficient for asymptotic stability of the equilibrium: (a) Outside of a certain spherical domain $|\mathsf{x}| < r$ an inequality of the form $v \geq s$ holds with constant s; (b) in the domain $v < s$, the derivative $\dot{v}$ is nonpositive; (c) in the domain $\dot{v} = 0$, there lies no complete half-trajectory of the differential equation with $t \geq t_0$ (Barbašin and Krasovskii [1], Tuzov [1]). The corresponding statement

for asymptotic stability in the whole can be formulated as an extension of Theorem 4.3.

8. In a criterion of Krasovskii [21], a family of Liapunov functions is used. This criterion is based on the fact that all motions originating in the interior of the domain of attraction of the equilibrium are asymptotically stable (cf. Sec. 2). With each "perturbed motion" $p(t, x_0, t_0)$, originating in the interior of the domain of attraction, a Liapunov function $v(x_0, t_0; x, t)$ is associated. The initial values are to be regarded as parameters. The function v is defined in a certain neighborhood

$$|x - p(t, x_0, t_0)| < \delta(x_0, t_0; t)$$

of the perturbed motion. The function δ is positive, continuous, and is permitted—this is essential—to tend towards zero with increasing t. Associated with this family of functions are four positive constants a_1, a_2, b, and c which are independent of the initial values, such that the inequalities

$$a_1|x|^b < v(x_0, t_0; x_0, t) < a_2|x|^b$$

$$\frac{d}{dt} v(x_0, t_0; x(t) - p(t, x_0; t_0), t) < -cv(x_0, t_0; x, t)$$

are valid. A solution different from p has to be substituted for $x(t)$ in the left hand side of the last inequality. The existence of such a family of functions is sufficient for the asymptotic stability of the equilibrium (even for exponential stability, cf. Sec. 22). Veksler [1] operated with a slightly different modification of the Liapunov function.

9. Makarov [1] considered two simultaneous equations

$$\dot{x} = f(x, y, t), \qquad \dot{y} = g(y, t) \qquad (4.7)$$

and called the unperturbed motion stable if

follows from $\qquad |y| < \epsilon_1 \quad \text{and} \quad |x - y| < \epsilon_2 \qquad (t > t_0)$

$\qquad\qquad\qquad |y| < \delta_1, \quad |x - y| < \delta_2 \quad \text{at} \quad t = t_0$

It is assumed that $g(0, t) \equiv 0$, but not necessarily $f(0, 0, t) \equiv 0$. Naturally, the equations could be discussed in a $(2n + 1)$-dimensional motion space by introducing a new variable $z = x - y$. However on the basis of the following theorem (the proof of which is similar to that of Theorem 4.1), it is possible to discuss Eq. (4.7) by means of two Liapunov functions (Makarov [1]): Let the function $v(x, y, t)$ be "positive definite for $x = y$" (that is, let $v(x, x, t) = 0$ and $v(x, y, t) = w(x, y)$, where $w > 0$ for $x \neq y$, and where $w = 0$ only for $x = y$). Furthermore, let $u(y, t)$ be positive definite. If the derivatives $\dot{v}$ and $\dot{u}$ for (4.7) are negative semi-definite, then the unperturbed motion is stable.

10. Basically, the application of the direct method is restricted to the real domain. If an equation is complex, it must be transformed into a system of real equations by splitting it into real and imaginary parts (cf. Sec. 2, Remark 3). This transformation may be formally bypassed by constructing a Liapunov function $v(z, \bar{z}, t)$ for the complex equation

$$\dot{z} = f(z, t)$$

with $v(\bar{z}, z, t) = \overline{v(z, \bar{z}, t)}$. (The bar denotes the conjugate complex.) In discussing such a function, separation into real and imaginary parts cannot be avoided.

11. The foregoing stability theorems have been formulated for the particular isolated singularity, the origin, corresponding to the equilibrium. Consequently, it was possible to characterize the Liapunov function v by the general properties "v is positive definite," or "v is decrescent," quite independently of the particular form of the equation. If it is desired to state the theorems of the direct method for an arbitrary solution rather than for the equilibrium, then the special properties of the solution will naturally enter the definition of v and the theorems lose their general character (cf. Sec. 20). However, if the point set $\dot{v} = 0$ is invariant, then it is sometimes possible to make a more precise statement. ($\mathfrak{M}$ is said to be *invariant* if each trajectory originating in $\mathfrak{M}$ remains in $\mathfrak{M}$ for all $t > t_0$; cf., e.g., Lefschetz [1].) LaSalle [1, 2] proved: Each solution of the autonomous equation $\dot{x} = f(x)$ originating in $\mathfrak{B}$ tends towards $\mathfrak{M}$ if the following three conditions are fulfilled: $\mathfrak{B}$ is of such a nature that each solution originating in $\mathfrak{B}$ is bounded and remains in $\mathfrak{B}$ for all $t > t_0 + T$ (T finite); $v(x) \in C_1$ and $\dot{v} \leq 0$ in $\mathfrak{B}$; $\mathfrak{M}$ is the greatest invariant subset of the set $\dot{v} = 0$ in $\mathfrak{B}$.

Example:

$$\dot{x} = (x - y)(1 - a^2x^2 - b^2y^2), \qquad \dot{y} = (x + y)(1 - a^2x^2 - b^2y^2)$$

We choose $v = x^2 + y^2$. Then

$$\dot{v} = (x^2 + y^2)(1 - a^2x^2 - b^2y^2)$$

Therefore, the phase curves approach the ellipse $a^2x^2 + b^2y^2 = 1$ from the outside and from the inside, depending on the sign of $\dot{v}$. The ellipse itself is not a limit cycle (it is not a solution), but it is an invariant set.

12. The main theorems can be extended to the case of stability with probability, mentioned in Sec. 2, Remark 6. In this case, the Liapunov function also depends on the random function appearing in the differential equation. Instead of the function $v(t)$, introduced in Sec. 1 (i), one has to operate with the mathematical expectation for that expression obtained by substituting a solution into v (Kats and Krasovskii [1]).

5. THEOREMS ON INSTABILITY

Compare the preliminary remarks to Sec. 4.

THEOREM 5.1 (Chetaev [1]): *Given the differential equation* (2.7) *and a function* $v(x, t)$ *with the following properties:*

(a) *In every domain* $\mathfrak{K}_\epsilon$ (*with arbitrarily small* $\epsilon > 0$) *there exist points* x *such that* $v(x, t)$ *is negative for all* $t \geq t_0$ (*with sufficiently large* t_0). *The totality of points* (x, t) *with* $|x| < h$ *and* $v(x, t) < 0$ *shall be denoted as the "domain* $v < 0$". *This domain is bounded by the hypersurfaces* $|x| = h$ *and* $v = 0$ *and is possibly separated into several subdomains* $\mathfrak{A}_1, \mathfrak{A}_2, \ldots$;

(b) *v is bounded below in a certain subdomain* $\mathfrak{A}$ *of the domain* $v < 0$;

(c) *In the domain* $\mathfrak{A}$ *of the motion space defined in* (b) *the total derivative* $\dot{v}$ *for* (2.7) *is negative; in particular,* $\dot{v} \leq -\varphi(|v|) < 0$, *where* $\varphi(r)$ *is a function of the class K. The existence of such a function* $v(x, t)$ *implies that the equilibrium is unstable.*

Proof: Let (x_0, t_0) be a point of the domain $\mathfrak{A}$ with arbitrarily small $|x_0|$ such that $v(x_0, t_0) = -\alpha < 0$, $\dot{v} < 0$. Along the motion $\mathsf{p}(t, x_0, t_0)$, v decreases. On the other hand [cf. (4.5)]

$$v(t) = v(\mathsf{p}(t, x_0, t_0), t) = -\alpha + \int_{t_0}^{t} \dot{v} \, d\tau < -\alpha - \varphi(\alpha)(t - t_0) \quad (5.1)$$

and, since v is bounded below in the domain under consideration, the motion must leave this domain. However, this can only occur at the boundary $|x| = h$; i.e., the equilibrium is unstable.

Theorem 5.1 contains as a special case the so-called *First Theorem of Liapunov on instability:*

THEOREM 5.2 (Liapunov [1]): *The equilibrium is unstable if there exists a decrescent function* $v(x, t)$ *which has a domain* $v < 0$, *and whose total derivative for* (2.7) *is negative definite.*

Corollary: If in this case v is always negative for $x \neq 0$, in particular if it is negative definite, the equilibrium is completely unstable (cf. Definition 2.6).

Another sufficient condition for instability is stated by the so-called *Second Theorem of Liapunov* [1] *on instability:*

THEOREM 5.3: *The equilibrium is unstable if the following holds: In the domain* $\mathfrak{K}_{h,t_0}$ *there exists a bounded function* $v(x, t)$ *with the properties:*

(a) *Its total derivative for* (2.7) *is of the form*

$$\dot{v} = gv + w(x, t) \quad (5.2)$$

where g is a positive constant, and where $w(x, t)$ *is a semi-definite function;*

(b) *If* $w(x, t)$ *does not vanish identically, there exist in each domain* $\mathfrak{K}_{h_1,t_1}$ *with arbitrarily large* t_1 *and arbitrarily small* $h_1 \leq h$ *such x-points that* $v(x, t)$ *and* $w(x, t)$ *have the same sign for* $t > t_1$.

Theorem 5.1 is generally more convenient for applications than Theorem 5.3. However, viewed from the standpoint of the theory, they are equivalent. As a matter of fact both furnish necessary and sufficient conditions (cf. Sec. 19): If the equilibrium is unstable there exist two functions v_1 and v_2 which satisfy the assumptions of Theorems 5.1 and 5.3, respectively. Consequently, the following weakenings of the assumptions are of interest only for the application of the theorems and do not amplify the region of their validity. According to Dubošin [7], one can replace in (5.2) the term gv by $g(t)v^k$, where $k \geq 1$ and where the integral

$$\int_{t_0}^{t} g(\tau)\, d\tau$$

increases beyond any bound. Erugin [4] stated several other assumptions weaker in a different sense. For instance, it suffices to require that for a certain unboundedly increasing sequence $t_1, t_2, \ldots$ of upper limits the integral

$$\int_{t_0}^{t_n} \dot{v}\, d\tau$$

decreases below any bound, or it even suffices that this integral for a fixed $t = t^*$ is less than the bound required in assumption (b) of Theorem 5.1 for the function v. Another modification of Theorem 5.1 was given by Massera [4].

Chetaev [2, 6] proved a theorem in which two functions are involved. Let it be possible to choose a decrescent function $v(x, t)$ and a function $w(x, t)$ in the following manner:

1. The domain $v\dot{v} > 0$ is not empty for any value of t in the interval $t_0 \leq t \leq \infty$ which has to be considered as a closed interval;

2. For arbitrarily small $|x|$ there exists a subdomain $w > 0$ of the domain $v\dot{v} > 0$; the derivative $\dot{w}$ has a constant sign on the boundary $w = 0$ of the domain $w > 0$.

Then the equilibrium is unstable.

In analogy to the theorems on stability mentioned at the end of Sec. 4.9, Makarov [1] established two theorems on instability in which the assumptions of Theorem 5.1 are required correspondingly for only one of the functions u and v, respectively.

The following theorem can be proved similarly to Theorem 5.1.

THEOREM 5.4 (S. K. Persidskii [1]): The equilibrium is completely unstable if a function $v(x, t)$ exists which has the following properties in a domain $\mathfrak{K}_{h,t_0}$:

1. $v > 0$ for $x \neq 0$;
2. $\dot{v} \geq 0$;
3. The function v tends uniformly towards zero as t increases.

The function v appearing in Theorem 5.4 does not have to be definite. The fact that one cannot dispense with the uniformity required under 3, is illustrated by the differential equation

$$\dot{x} = -x/4$$

whose equilibrium is asymptotically stable. The functions
$$v = x^2 \exp(t - t^2 x^4), \qquad \dot{v} = (\tfrac{1}{2} - 2tx^4 + t^2 x^4)v$$
satisfy for $t_0 \geq 2$ all assumptions of the theorem, except that of uniformity. The same example proves, by the way, that the boundedness of v in $\mathfrak{K}_{h,t_0}$, required by Theorem 5.3, must not be dropped.

3 APPLICATIONS OF THE STABILITY THEOREMS TO CONCRETE PROBLEMS

6. FUNDAMENTAL REMARKS ON THE APPLICATIONS

In many problems, statements on the stability properties of an equilibrium can be made by means of the theorems of Chap. 2. However, the method has to be adjusted to the peculiarities of every specific system. There is no general procedure for the construction of a Liapunov function. (Cf. Chap. 4 for the problem of the existence of Liapunov functions.) In particular the following problems can be investigated:

1. *Clarification of the question of whether an equilibrium is stable or unstable.* If the stability of a motion different from the equilibrium is to be investigated, the differential equation of the perturbed motion (cf. Sec. 2) must first be established (cf. Ergen, Lipkin and Nohel [1]).

2. *Estimates of the stability domain of the initial points.* The domain contains all of those points x of the phase space from which motions originate which either remain in a prescribed neighborhood of the origin in the case of weak stability, or which come arbitrarily close to the origin in the case of asymptotic stability. That part of the stability domain for which the latter holds is (as mentioned in Sec. 2) the domain of attraction of the equilibrium (or of the trivial solution).

3. *Estimates of the stability domain of the parameters.* Frequently, the equations of motions, particularly equations of physical systems, depend on parameters which vary between certain bounds. The question of interest is for which of the parameter values the equilibrium will be stable. It is convenient to interpret these parameters as coordinates of a parameter space. The subspace of all "points" for which the equilibrium is asymp-

totically stable is called the *stability domain of the parameters*. In most physical and technical applications, asymptotic stability is required even in the whole.

4. *Derivation of conditions for nonlinearities in order to guarantee stability*. The nonlinear parts of the differential equations must have certain properties, or satisfy certain conditions for the equilibrium to be asymptotically stable. For instance, these nonlinearities must have certain properties of rate of change; their maxima or minima must lie between certain bounds; certain integral inequalities must be satisfied; etc. A clear separation of these problems and those mentioned under 3 is not always possible.

5. *Estimates for the solutions*. As can be seen from the geometrical interpretation of the direct method sketched in Sec. 3, a relation exists between the absolute values of the solutions and the values of the Liapunov function. Therefore, estimates of the integrals by means of the Liapunov functions can be given. However, inequalities of this nature are of value only if the Liapunov function is of a sufficiently simple form. Sometimes it is sufficient to know the behavior of the Liapunov function only in certain domains of the motion space which do not necessarily have to be neighborhoods of the t-axis (Melnikov [1]).

The problems mentioned under 2 to 4 are usually treated as follows: We construct a Liapunov function, and we investigate for which initial points, or for which points of the parameter space, etc., this Liapunov function looses its characteristic properties. In general, no "optimal" results can be obtained along this line, since the theorems of Sec. 4 furnish only sufficient conditions. However, it has to be emphasized that some stability problems can as yet be investigated only by means of the direct method (Secs. 13 and 14).

The applications of the direct method known at this time refer almost exclusively to autonomous differential equations. This is not only a consequence of the difficulties connected with nonautonomous differential equations, but it is also a consequence of the fact that the equations of motion of most concrete systems are autonomous.

6. Krasovskii [25, 26] pointed out another rather specialized application. In the field of *optimal control*, we are concerned with the problem of designing the control process so that the time of the control action becomes a minimum. In the linear case, let the system be governed by the equation,

$$\dot{x} = Ax + Bu + cy$$

where $u = u(t)$ is the control vector, and where $y = y(t)$ represents a scalar perturbation function. The motion depends on t, x_0, and, in addition, on y, $y_0 = y(t_0)$ and u. If we consider the time T^0 of the control action along a certain trajectory, it turns out that the derivative dT^0/dt is equal to -1 along an optimal trajectory. Considered as a function of the initial values x_0 and t_0, T^0 is basically positive definite, and, as a consequence of $dT^0/dt < 0$, the assumptions of Theorem 4.2 are satisfied with $v = T^0$. Hence, the problem of optimizing the process can be considered as the problem of constructing a Liapunov function with the additional property $\dot{v} = -1$. Krasovskii applied this consideration in order to prove the existence of solutions of the optimization problem.

7. EQUATIONS WITH DEFINITE FIRST INTEGRALS

A definite first integral $v = $ const (if it exists) of Eq. (2.7) can be used as a Liapunov function. In this case, $\dot{v}$ vanishes identically. Applying Theorem 4.1, we can conclude that the equilibrium is stable.

For example, let the equations of motion of an autonomous system be given in canonical form:

$$\dot{q}_i = \partial H/\partial p_i, \quad \dot{p}_i = -\partial H/\partial q_i \quad (i = 1, \ldots, n) \quad (7.1)$$

These equations have the energy integral $H = $ const. The kinetic energy is a positive definite form of the momenta p_i. If the potential energy $U(q_1, \ldots, q_n)$ takes on a minimum at the equilibrium $q_1 = \cdots = q_n = 0$, which can always be assumed to be 0, then the total energy is positive definite. Theorem 4.1 then leads to the well known *Theorem of Lagrange*: The equilibrium is stable if the potential energy takes on a minimum at this position (cf. Sec. 26 and Pozarickii [1]).

If H is not definite, one can try to construct a definite function $v = H + w$ by adding a function $w(p, q)$. The total derivative of this function for (7.1) is equal to the negative Poisson expression.

$$\dot{v} = -(w, H) = -\sum_{i=1}^n \left(\frac{\partial w}{\partial p_i} \frac{\partial H}{\partial q_i} - \frac{\partial w}{\partial q_i} \frac{\partial H}{\partial p_i} \right)$$

One has to investigate the question whether or not this expression is definite (Liapunov [1], Dubošin [2]). If the potential energy is a homogeneous function of the coordinates q_i which takes on negative values in every neighborhood of the origin, then the equilibrium is unstable. This can be shown by means of the Liapunov function

$$v = H \sum_{i=1}^n p_i q_i$$

and by applying Theorem 5.1 (Chetaev [8]).

The following theorem answers the question under what conditions a definite first integral can be constructed at all from first integrals of the equation of motion.

THEOREM 7.1 (Pozarickii [3]): Let $u_1(x, t), \ldots, u_p(x, t)$ $(p < n)$ be p known linearly independent first integrals of $\dot{x} = f(x, t)$. A necessary and sufficient condition for the existence of a definite first integral of the form $\varphi(u_1, \ldots, u_p)$ is the definiteness of the form $u_1^2 + \cdots + u_p^2$.

Naturally, this form is always semi-definite. By means of Theorem 7.1, Pozarickii showed, for instance: No definite first integral can be constructed from $u_1, \ldots, u_p$ if the known integrals do not depend explicitly on t, and if they are of the form

$$u_i = a_i^T x + \text{terms of higher order}$$

and if the rank of the matrix $(a_1, \ldots, a_p)$ is equal to p.

The theory of gyros is a special field for applications of such considerations. In the simplest case of a gyro rotating about its center of gravity, the equations of motions are of the form

$$a\dot{p} + (c-b)qr = 0, \qquad b\dot{q} + (a-c)pr = 0,$$
$$c\dot{r} + (b-a)pq = 0$$

a, b, c are the principal moments of inertia; p, q, r are the components of the velocity vector in the coordinate system of the principal axes of inertia. In order to investigate the stability properties of the specific motion

$$p = 0, \quad q = 0, \quad r = \hat{r} \neq 0$$

we consider the differential equation of the perturbed motion, which is obtained by the substitution

$$x_1 = p, \quad x_2 = q, \quad x_3 = r - \hat{r}$$

This equation has the two integrals,

$$\frac{a-c}{b} x_1^2 + \frac{b-c}{a} x_2^2 \pm (ax_1^2 + bx_2^2 + 2c\hat{r}x_3 + cx_3^2)^2$$

which are definite for $a \geq b > c$ and $a \leq b < c$. We conclude that the rotations about the maximum and the minimum principal axes are stable. The instability of the rotation about the mean axis follows from Theorem 5.1, if we use the Liapunov function $v = x_1 x_2$ (Chetaev [7, 8]).

For a general motion of a gyro, the first integrals furnished by the theory, are in general not definite. However, under certain conditions a definite integral can be obtained from these integrals by suitable combinations of them. If this is possible, stability conditions can be established. Chetaev [7, 8] and Skimel' [1] proceeded in this manner in the case of the ordinary gyro, Rumiancev [1, 5], Zhak [1] and Charlamov [1] did so in the case of the motion of a gyro in a liquid medium, Beleckii [1] in the case of the motion of a gyro in a Newtonian field of force. (cf. also Morosova [1].)

By means of the method of definite first integrals, Magnus [2] investigated thoroughly the stability properties of the heavy symmetric gyro in gimbled suspension. Rumiancev [7] and Chetaev [9] used the paper of Magnus as a starting point. For further applications to the theory of gyros, the reader is referred to Czan Sy-Ni [2], Isaeva [1], Krementulo [1, 2], Pirogov [1], Rumiancev [8 to 12], Tabarovskii [1, 2].

A more general problem was investigated by Aminov [1]. He considered the paths of mass points, namely the geodetic curves in a Riemannian manifold, determined by the equation

$$\ddot{q}^j + \Gamma^j_{\alpha\beta}\dot{q}^\alpha \dot{q}^\beta = 0$$

Here, as usual in differential geometry, the dot denotes the derivative with respect to arc length. The "stability" of the transition to the "perturbed" trajectories

$$\hat{q}^j = q^j + p^j$$

is investigated. Under certain assumptions about the dependence of the metric fundamental form upon the perturbations p^j, first integrals of the equations can be found, which might be used to establish stability conditions.

8. CONSTRUCTION OF A LIAPUNOV FUNCTION FOR A LINEAR EQUATION WITH CONSTANT COEFFICIENTS

Let

$$\dot{x} = Ax \tag{8.1}$$

be the equation of motion, and let $p_1, \ldots, p_n$ be the eigenvalues of the matrix $A = (a_{ik})$. The Liapunov function shall be determined as a quadratic form,

$$v = x^T B x \quad (B^T = B) \tag{8.2}$$

under the condition that its total derivative for (8.1) is equal to a given negative definite quadratic form $-x^T C x$. This leads to the equation

$$A^T B + BA = -C \quad \text{(C positive definite)} \tag{8.3}$$

for the unknown matrix B. This equation represents a system of linear equations for the $n(n + 1)/2$ elements b_{ik} of B. This system of linear equations has a unique solution if none of the numbers p_i and none of the sums $p_i + p_j$ $(i, j = 1, \ldots, n)$ vanish. The matrix B determined in this manner is the matrix of a positive definite quadratic form (8.2) if and only if all eigenvalues p_i have negative real parts:

$$\operatorname{Re} p_i < 0 \quad (i = 1, \ldots, n) \tag{8.4}$$

(proof, cf. e.g. Hahn [3], Bellman [1], Taussky [1]). If condition (8.4) is satisfied, a Liapunov function is found which satisfies the assumptions of Theorem 4.2.

THEOREM 8.1: *The equilibrium of equation* (8.1) *is asymptotically stable if all eigenvalues of the matrix* A *have negative real parts.*

If there are eigenvalues with positive real parts, the solution B of (8.3)—

if it exists—is a matrix of an indefinite form, or the matrix of a negative definite form if $\operatorname{Re} p_i > 0$ ($i = 1, \ldots, n$); then the function (8.2) satisfies the conditions of Theorem 5.2. If Eq. (8.3) has no solution because one of the terms $p_i + p_j$ vanishes, then there exists always a number $q > 0$ such that the matrix $A_1 = A - qU$ still has eigenvalues with positive real parts, but no longer belongs to the exceptional case. Then B can be computed from the equation

$$A_1^T B + B A_1 = -C \quad (\text{C positive definite}) \tag{8.5}$$

The Liapunov function constructed according to (8.2) satisfies the assumptions of Theorem 5.3. If, in addition, the Corollary of Theorem 5.2 is used, the following theorem results:

THEOREM 8.2: *The equilibrium of Eq.* (8.1) *is unstable, if at least one of the eigenvalues of* A *has a positive real part. If the real parts of all eigenvalues are positive, then the equilibrium is completely unstable.*

A differential equation of the form (8.1), satisfying the conditions of Theorem 8.1, or of Theorem 8.2, shall in the sequel be called an *equation with significant behavior* of the equilibrium. In this case, the matrix has either only eigenvalues with negative real parts, or has at least one eigenvalue with positive real part. For a differential equation with *critical behavior* of the equilibrium, none of the eigenvalues has a positive real part; however, eigenvalues with vanishing real parts do occur. As we shall see later (Sec. 22), significant behavior of the equilibrium actually means that the solutions of the differential equation can be estimated exponentially. With suitable positive constants, we either have

$$|p(t, x_0, t_0)| \leq a|x_0|e^{-\alpha(t-t_0)} \tag{8.6}$$

for all solutions, or there are solutions with

$$|p(t, x_0, t_0)| \geq a|x_0|e^{+\beta(t-t_0)} \tag{8.7}$$

On the other hand, in a critical case, the not decaying solutions are either bounded, or they increase at most like a power of t.

The following procedure can be used to study the stability behavior in a critical case (Hahn [3]). Let $z = Lx$ be a nonsingular transformation, where L transforms A into its Jordan Normal form, $LAL^I = J$. The differential equation

$$\dot{z} = Jz \tag{8.8}$$

and differential equation (8.1) are equivalent with regard to the stability behavior of their equilibria. In general, differential equation (8.8) splits into several subsystems of equations of lower dimension. Theorem 8.1 holds for those subsystems associated with eigenvalues with negative real parts. If the elementary divisor which belongs to an eigenvalue with vanishing real part is linear, then the corresponding subsystem can be transformed into the form

$$\dot{z}_1 = 0, \quad \text{or} \quad \dot{z}_1 = kz_2, \quad \dot{z}_2 = -kz_1 \tag{8.9}$$

respectively (k real). If we take $v = z_1^2$ and $v = z_1^2 + z_2^2$, respectively, as Liapunov functions, then in both cases, $\dot{v} \equiv 0$ and Theorem 4.1 can be applied. If there are elementary divisors of second order, then there are subsystems with coefficient matrices

$$\begin{pmatrix} 0 & 1 \\ 0 & 0 \end{pmatrix} \quad \text{or} \quad \begin{pmatrix} 0 & -k & +1 & 0 \\ +k & 0 & 0 & +1 \\ 0 & 0 & 0 & -k \\ 0 & 0 & +k & 0 \end{pmatrix}$$

respectively. Here, we take $v = z_1 z_2$ with $\dot{v} = z_2^2$ and $v = z_1 z_3 + z_2 z_4$ with $\dot{v} = z_3^2 + z_4^2$, respectively, as Liapunov functions. By means of Theorem 5.1, or by means of the theorem which we obtain from 5.1, by replacing v by $-v$ (cf. the introductory remark of Sec. 4), we conclude that the system is unstable. This result remains valid for elementary divisors of higher order, and proves:

THEOREM 8.3: *The equilibrium of Eq.* (8.1) *is weakly stable if in a critical case the elementary divisors corresponding to eigenvalues with vanishing real parts are linear. The equilibrium is unstable if there are elementary divisors of higher order.*

Remark: The proof just sketched does not exclude asymptotic stability of the equilibrium. The more precise statement "weakly stable" which can be concluded immediately from the well-known explicit form of the solutions in terms of exponential functions, can also be obtained by means of the direct method. However, for that purpose we have to conclude, from the more complicated converse theorems of Sec. 18, that the equilibrium can never be asymptotically stable if there is no positive definite Liapunov function with negative definite derivative. But as it is readily seen, the derivative of a positive definite function for (8.9) can never be negative definite. It is sufficient to consider the quadratic terms of the power series expansion of v, which certainly exists (cf. Sec. 18).

Theorems 8.1 to 8.3 remain valid for a matrix with complex coefficients. This can be proved either by means of an Hermitian form, constructed in analogy to (8.2), or by reducing the complex system to systems with real coefficients (Vejvoda [1], Hahn [7]).

The elements b_{ik} of the matrix B defined by (8.3), can be expressed in closed form. For $\mathsf{C} = 2\mathsf{U}$, for example, we have

$$b_{ik} = \sum_{r,s}^{n} \frac{d_{ir}(p_s) d_{kr}(-p_s)}{d'(p_s) d(-p_s)} \tag{8.10}$$

where $d(p)$ denotes the characteristic polynomial of the matrix A whose zeros p_s are assumed to be simple. The quantities $d_{ik}(p)$ are the cofactors of the elements of the matrix $A - pU$. Equation (8.10) can be obtained by means of a formula of Malkin [15] (cf. Theorem 24.5) which is based on an investigation of Bedel'baev [1]. From Eq. (8.10) Bedel'baev derived some other formulae for the computation of the b_{ik}.

If for $\dot{v}$ the special form $\Sigma\, x_i x_k$ ($i \leq k$) is chosen, then v can be computed by an elementary procedure given by Moisseev (Bedel'baev [2]).

Instead of the quadratic form (8.2), a form of higher order can be used as a Liapunov function for (8.1). In this case, the following theorems of Liapunov [1], which refer to a differential equation of the form (8.1), hold.

THEOREM 8.4: If all Re $p_i < 0$, then there exists exactly one positive definite form v of the order m whose derivative $\dot{v}$ for (8.1) is equal to an arbitrarily given negative definite form of the order m.

THEOREM 8.5: If there is at least one Re $p_j > 0$, then, for a given form w of the order m, a form v of the same order and a number $q > 0$ can be found such that $\dot{v} = qv + w$ is the total derivative of v for (8.1) and such that the conditions of Theorem 5.3 are satisfied.

Disregarding certain exceptional cases, a form of the order m can also be constructed under the conditions of Theorem 8.5 such that the conditions of Theorem 5.2 are satisfied.

Liapunov proved Theorems 8.1 to 8.5 by using the stability behavior of the equilibrium of differential equation (8.1), which can be derived from the explicit form of the solutions. This means that Liapunov did not prove the statements about the equilibrium of (8.1) by means of the direct method.

9. SIMPLE STABILITY CONSIDERATIONS FOR NONAUTONOMOUS LINEAR DIFFERENTIAL EQUATIONS

The stability theory of general linear differential equations requires, in spite of their relatively simple form, a more profound approach; in particular, the Liapunov theory of the characteristic numbers (cf. Sec. 25) is required. But if the matrix $A(t)$ of the equation

$$\dot{x} = A(t)\, x \tag{9.1}$$

is bounded, and if the eigenvalues of $A(t)$ have negative real parts for every fixed $t \geq t_0$, then some sufficient stability conditions can be obtained by modifying the methods developed in Sec. 8.

We compute the symmetric matrix $B(t)$ which depends on t from the equation

$$A^T(t)\, B(t) + B(t)\, A(t) = -C \tag{9.2}$$

This equation is analogous to Eq. (8.3). If the given constant symmetric matrix C is positive definite, then, as a consequence of the assumptions on A, B(t) is uniquely determined for fixed t and B(t) has positive eigenvalues. Let $\varphi(t)$ for $t \geq t_0$ be a positive bounded function whose derivative is continuous. As a Liapunov function for (9.1), we choose the quadratic form

$$v = \varphi(t)\, x^T B(t)\, x \tag{9.3}$$

The total derivative of this form for (9.1) is

$$\dot{v} = \varphi(t)\, x^T G(t)\, x$$

where

$$G(t) = -C + \dot{B}(t) + \frac{d \log \varphi(t)}{dt} B(t)$$

Let $\mu(t)$ be the greatest root of the equation

$$\det(\dot{B} - C - \mu B) = 0 \tag{9.4}$$

and let

$$\gamma(t) = \varphi(t_0) \exp\left(-\int_{t_0}^{t} \mu(\tau)\, d\tau\right) \tag{9.5}$$

Then the condition

$$\frac{d \log \varphi}{dt} < \frac{d \log \gamma}{dt} = -\mu$$

is necessary and sufficient for G(t) to have only negative eigenvalues. In this case, $\dot{v}$ is negative. Now, the function $\varphi(t)$ has to be chosen such that v is positive definite. According to Lebedev [3], it is sufficient for this purpose, for example, that the expression $\varphi(t) \sqrt{\beta(t)}$ be always greater than a fixed positive number δ; $\beta(t)$ here denotes the smallest eigenvalue of B(t). As a *sufficient condition* for the stability of the equilibrium, the following inequality is obtained (Lebedev [3]):

$$\sqrt{\beta(t)} \exp\left(-\int_{t_0}^{t} \mu(\tau)\, d\tau\right) \geq \delta > 0 \qquad (t \geq t_0) \tag{9.6}$$

Asymptotic stability cannot readily be inferred, since the definiteness of $\dot{v}$ is not secured. If the real parts of the eigenvalues of A(t) are always smaller than a negative bound which is independent of t, then the form (9.3) is positive definite even for $\varphi(t) \equiv 1$.

In a similar way, a stability condition can be established for a differential equation with "slowly varying coefficients." For these equations, the matrix $\dot{A}(t)$ is equal to a matrix $\gamma Q(t)$ with bounded Q(t). If B is determined according to (9.2), then $\dot{B} = \gamma R(t)$, where R(t) is bounded, and for $v = x^T B x$, we obtain

$$\dot{v} = -x^T(C - \gamma R(t))\, x$$

(Razumichin [5]). The condition that $\dot{v}$ be negative semi-definite, or negative definite, respectively, furnishes bounds for γ. However, without additional assumptions on $A(t)$, again only weak stability can be inferred (Chetaev [2, 3]).

Sometimes it is convenient to introduce a new time scale. Let the real parts of the eigenvalues of $A(t)$ be always smaller than a given negative number. Furthermore, let τ be a new variable and let

$$t = t(\tau) \qquad (t_0 = t(0), \tau \geq 0)$$

be a continuous and continuously differentiable function. The derivative with respect to τ (denoted by a prime) shall always be bounded by two fixed positive numbers:

$$0 < k_1 < t'(\tau) < k_2$$

The original equation (9.1) is transformed into

$$x' = t'(\tau) A(t(\tau)) x \tag{9.7}$$

and the matrix $B(t)$ is replaced by the matrix

$$B^*(\tau) = (1/t') B(t(\tau))$$

The total derivative of the Liapunov function $v^* = x^T B^* x$ for (9.7) is

$$v^{*\prime} = x^T (\dot{B}(t) - B(t)\dot{s} - C) x$$

Here, $\log t' = s$, and this quantity has to be considered as a function of t. According to the conditions on t', these expressions always make sense. We try to make $v^{*\prime}$ negative definite by choosing $\dot{s}(t)$ sufficiently large. We choose, for example, $\dot{s}(t)$ greater than the greatest root $\tilde{\mu}(t)$ of the equation

$$\det(\dot{B}(t) - \mu B(t) - C) = 0$$

which is formally equivalent to (9.4). We then have

$$t' = e^s = \exp\left(\int_{t_0}^t \dot{s}\, dt\right) \geq \exp\left(\int_{t_0}^t \tilde{\mu}(\tau)\, d\tau\right)$$

Since t' is bounded, we obtain the sufficient stability condition given by Razumichin [3] in the form:

$$\int_{t_0}^t \tilde{\mu}(\tau)\, d\tau \tag{9.8}$$

must be bounded.

Let the matrix $A(t)$ in (9.1) be a periodic function with period ω and let

$$C = \frac{1}{\omega} \int_0^\omega A(t)\, dt$$

be similar to a diagonal matrix D. If the original equation is transformed by means of a linear transformation into the form

$$\dot{y} = Dy + F(t) y$$

then the discussion of the Hermitian form

$$v = y^T(D + \bar{D})\, y$$

which has to be used as a Liapunov function, leads to the following results (Nougmanova [1]):

(a) The conditions
   D negative definite,

$$G = (D + \bar{D})^2 + (D + \bar{D})\, F + F^T(D + \bar{D})$$

positive, are sufficient for stability;

(b) The conditions
   D negative definite,
   G positive definite,
   are sufficient for asymptotic stability.

By means of simple Liapunov functions, Dubošin [3] obtained sufficient stability conditions for the special differential equation

$$\ddot{x} + p(t)\, x = 0$$

which is equivalent to the system

$$\dot{x} = y, \qquad \dot{y} = -p(t)\, x$$

This system has already been investigated by Liapunov. For instance, let $p(t)$ be negative for large t; then the equilibrium is unstable. This follows from Theorem 5.1 by means of the Liapunov function $v = xy$. If $\lim p(t)$ exists as $t \to \infty$, if this limit value is positive, and if $\dot{p}(t) < 0$, then the equilibrium is asymptotically stable. In this case, $v = (px^2 + y^2)\, e^p$ is chosen as a Liapunov function.

Reissig [7, 8,] used Remark 4 of Sec. 4 for the construction of a Liapunov function for a system of second order. Other references are: Charasachal [2] and Reissig's [5] comment on this note.

Roitenberg [1] considered a system of linear differential equations of higher than first order in the form

$$\sum_{k=1}^{n} f_{jk}(D)\, x_k = \sum_{k=1}^{n} l_{jk}(D)\, x_k \qquad (j = 1, \ldots, n;\, D = d/dt)$$

Here, the f_{jk} are certain polynomials with constant coefficients, the l_{jk} are polynomials with variable coefficients. By means of a suitable linear transformation, the system can be brought into the form

$$\dot{y} = (J + B(t))\, y$$

The Jordan matrix J does not depend on t. The transformation formulae are given explicitly. Application of Theorem 4.2 with the Liapunov function $v = |y|^2$ furnishes sufficient stability conditions.

10. EQUATIONS WITH LINEAR PRINCIPAL PARTS

Let
$$\dot{x} = Ax \qquad (10.1)$$
be an autonomous linear differential equation with significant stability behavior (cf. Sec. 8). Suppose v^0 is a Liapunov function for (10.1), constructed on the basis of the theorems of Sec. 8, and suppose v^0 is a form of the order m. In practical applications we usually assume $m = 2$. The derivative $\dot{v}^0$ is then definite. Only in the exceptional case where $p_i + p_j = 0$, the function $\dot{v}^0$ is of the form $qv +$ definite form $(q > 0)$.

In addition to Eq. (10.1), we consider the differential equation
$$\dot{x} = Ax + g(x) \qquad (10.2)$$
which is in general nonlinear. We shall say that Eq. (10.1) is obtained from the Eq. (10.2) by *reduction*, i.e., by neglecting the term $g(x)$. We shall try to find a Liapunov function for (10.2) of the form
$$v = v^0 + v^z$$
where the additional term v^z has to be so chosen that:

1. v has the same definiteness properties as v^0,
2. $\dot{v}_{(10.2)}$ has the same definiteness properties as $\dot{v}^0_{(10.1)}$.

Here, the total derivative of v for (10.2) is
$$\dot{v}_{(10.2)} = \dot{v}^0_{(10.1)} + \sum_{i=1}^{n} \frac{\partial v^0}{\partial x_i} g_i + \dot{v}^z_{(10.1)} + \sum_{i=1}^{n} \frac{\partial v^z}{\partial x_i} g_i \qquad (10.3)$$

If a function v^z can be chosen such that v satisfies the above mentioned conditions, then the differential equations (10.1) and (10.2) are equivalent with respect to the stability behavior of their equilibria. In the construction of v^z, the specific properties of $g(x)$ have to be considered.

Special cases:

(a) Suppose the functions $g_i(x)$ have, in the neighborhood of the origin, power series expansions in the variables $x_1, \ldots, x_n$ beginning with terms of at least second order. Then Eq. (10.1) is called the *equation of the first approximation* of Eq. (10.2) and we shall say that (10.1) is obtained from (10.2) by *linearization*. In this case, we can use the function $v^z \equiv 0$, since the second term of the right hand side of (10.3) is at least of the order $m + 1$ in the variables x_i. Therefore, this term has no effect on the definiteness behavior of the term $\dot{v}^0_{(10.1)}$. This leads to the fundamental *Theorem on the Stability in the First Approximation.*

THEOREM 10.1 (Liapunov [1]): *If the stability behavior of the differential equation of the first approximation is significant, then the equilibrium of the complete differential equation has the same stability behavior as the equilibrium of the reduced equation.*

This theorem can be proved also by other means (cf., for instance, Perron [1] with weaker assumptions on $g(x)$, or $g(x, t)$, respectively; compare (b) and (c) below). However, the proof recently given by Saltykow [1] is also based on the construction of a Liapunov function.

In critical cases, the stability behavior of the equilibrium is not determined only by the first order terms. In fact, as Liapunov [1] showed, for a fixed linear part instability or asymptotic stability of the equilibrium can be obtained by modifying the terms of higher order.

(b) The assumption about $g(x)$ made under (a) is very restrictive and is not satisfied in many practical problems. Therefore, we replace it by the following weaker assumption: Every function $g_i(x)$ can be estimated by two fixed linear forms, i.e., there are vectors g'_i, g''_i ($i = 1, \ldots, n$) such that

$$x^T g'_i \leq g_i(x) \leq x^T g''_i \quad \text{or} \quad x^T g'_i \geq g_i(x) \geq x^T g''_i \quad (i = 1, \ldots, n) \quad (10.4)$$

We further assume that all eigenvalues $p_1, \ldots, p_n$ of the matrix A have negative real parts such that the equilibrium of the reduced differential equation is asymptotically stable. The problem is to find bounds of the components g'_{ik}, g''_{ik} of the vectors g'_i, g''_i such that the inequalities (10.4) guarantee the asymptotic stability of the equilibrium of (10.2). To solve this problem, we consider in addition to (10.1) the linear auxilliary equation

$$\dot{x} = (A + G) x \quad (10.5)$$

in which the constant matrix G is still to be determined. By forming the total derivative of the quadratic form (8.2) for (10.5), we obtain a quadratic form with the matrix

$$M = -C + G^T B + BG \quad (10.6)$$

We choose G such that M becomes negative definite. Then, as a consequence of Theorem 4.2, the equilibrium of (10.5) is asymptotically stable. The conditions for definiteness of M furnish n inequalities for the n^2 elements of the matrix G. The common solutions of these inequalities determine in the space of the g_{ik} a well-defined domain which is certainly not empty, since it contains a neighborhood of the origin. If the numbers g'_{ik}, g''_{ik} lie in that interval in which g_{ik} may vary without leaving the above mentioned domain, then the function (8.2) is a Liapunov function for the differential equation (10.2), provided condition (10.4) is satisfied (Aizerman [4], Hahn [2]). The degree of the inequalities which have to be solved in order to determine the elements of G is at most $2r$, where r is the number of nonvanishing components of the vector $g(x)$ (Hahn [2]). In practical problems, sometimes only one of the scalar equations which correspond to the vector equation (10.2) is actually nonlinear. In this case, the inequalities are quadratic. But even then, the computations are relatively cumbersome (Aizerman [4] and Pestel [1]).

Estimates of the form (10.4) can certainly be given if $|g(x)| < a |x|$ with $a > 0$. In this case, the theorem on the stability of the first approximation (for sufficiently small a) is also valid (Perron [1]). Evidently, this method does not give very accurate results. However, the results are in general not worse than those obtained by other methods. The accuracy of the results depends heavily on the properties of the linear part of the differential equation (10.2), as shown by Malkin [17].

With respect to the solution of the problem mentioned in connection with (10.4), the matrix C can be chosen arbitrarily. Lehnigk [1, 2] showed by investigating some special systems that the "quality" of the results, in other words, that the bounds for the elements of G, depend very much on C. C = U is not always the best choice, although it is particularly convenient for computations.

(c) The considerations under (b) remain valid if the right hand side of (10.2) depends explicitly on t, provided the linear estimates (10.4) hold for $t \geq t_0$ uniformly in t (Perron [1]). For example, this condition is satisfied if the original equation is of the form

$$\dot{x} = (A + Q(t)) \qquad (10.7)$$

in which the matrix A is constant and the matrix $Q(t)$ has bounded elements which approach zero as t increases. The stability behavior of the equilibrium of Eq. (10.7) is determined by the *limit equation* (8.1), provided that this limit equation has significant behavior of the equilibrium (Chetaev [8]). This statement can be expressed in a more general form (cf. Sec. 26, Remark to Theorem 26.2).

Erugin [1] considered an equation where the matrix of the linear part is of the form $A + \epsilon F(t)$. The elements of $F(t)$ are trigonometric polynomials; ϵ is a small parameter. The considerations under (a) and (b) are applied twice. In the first step, bounds for the parameters are obtained; in the second step, bounds for the nonlinear terms are obtained.

Chetaev [11] considered the equation $\dot{x} = Ax + \epsilon f(x, t)$ and obtained bounds for ϵ which guarantee asymptotic stability.

In certain cases, the methods just sketched can also be extended to differential equations with discontinuous right hand sides, in particular to equations with terms of the type sgn x (Aizerman and Gantmakher [2], Anosov [1], Litvartovskii [1, 2, 3]). A report on differential equations of such a nature was published by Filippov [1].

11. BOUNDS FOR THE INITIAL VALUES

Suppose the equilibrium of the linear part (10.1) of differential equation (10.2) is asymptotically stable. We consider the Liapunov function v, constructed for the linear part, and its derivative $w(x)$ constructed for (10.2). Let the function v be radially unbounded, which, for example, is the case for (8.2), or, in general, for every definite form. If $g \in E$ holds everywhere, and if $w(x)$ is negative definite throughout the entire phase space, then, as a consequence of Theorem 4.3, the equilibrium is asymptotically stable in the whole. However, if $w(x)$ vanishes for certain points $x \neq 0$, we cannot draw this conclusion. But as long as one of the hypersurfaces $v(x) =$ const lies completely in the interior of the domain determined by $w(x) = 0$

(cf. Sec. 3, Remark 2), then it certainly belongs to the domain of attraction of the origin. Hence, it is necessary to determine those surfaces $v = $ const which contact the surface $w = 0$ from the inside, and to select from among all these surfaces that one which corresponds to the smallest value v_0 of the constant. Then the domain

$$v(\mathsf{x}) < v_0 \tag{11.1}$$

is a subdomain of the domain of attraction. At a point y at which those surfaces are in contact with each other, grad v is parallel to grad w. Therefore,

$$w(\mathsf{y}) = 0, \quad \text{and rank} \; \left.\frac{\partial(v, w)}{\partial(x_1, \ldots, x_n)}\right|_{\mathsf{x}=\mathsf{y}} = 1 \tag{11.2}$$

Certain liberty is involved in the construction of the Liapunov function. For example, the matrix C in (8.3) may be given arbitrarily as a symmetric positive definite matrix. Therefore, it is possible to vary the domain (11.1) in a certain manner by a suitable choice of v. Kunin [1] and Kartvelishvili [1] investigated a practical problem of obtaining an optimum in the case of $n = 2$. This problem deals with the stability of the motion of water in a storage basin of a power plant.

The considerations of 10(b) and 10(c), respectively, have to be modified if the linear estimates (10.4) do not hold for all x. For example, let (10.4) be valid only in the interior of the n-dimensional parallelepipedon $|x_i| \leq a_i$, whereas for $|x_i| > a_i$ the inequalities be violated. We then consider those surfaces $v = $ const which contact the hyperplanes $x_i = \pm a_i$ (the corresponding conditions are obtained in analogy to (11.2)), and we again select that surface with the smallest value of the constant. Then an inequality analogous to (11.1) determines a subdomain of the domain of attraction.

If the inequalities (10.4) hold only for large values of $|\mathsf{x}|$, for example for $|x_i| > a_i$, then as a consequence of the method of construction, the function is negative with certainty only for large values of $|\mathsf{x}|$. In this case, we can infer only that phase curves originating from large initial points approach a neighborhood of the origin in which they remain. A simple adjustment of the considerations carried out above leads to an estimate of this neighborhood, and hence to bounds for the motion itself (Aizerman [4], Hahn [4]).

If we consider (8.1) as a special case of a nonlinear system

$$\dot{\mathsf{x}} = \mathsf{f}(\mathsf{x}) \qquad (\mathsf{f} \in C_1) \tag{11.3}$$

then $\mathsf{Ax} = \mathsf{f}(\mathsf{x})$, and A is equal to the functional matrix

$$\mathsf{J}(\mathsf{x}) = \left(\frac{\partial(f_1, \ldots, f_n)}{\partial(x_1, \ldots, x_n)}\right)$$

at the point $\mathsf{x} = 0$. Theorem 8.1 may be considered as a stability statement based on a property of this functional matrix. A logical generalization is given by:

THEOREM 11.1 (Krasovskii [9, 10, 21]): The equilibrium of $\dot{x} = f(x, t)$ is asymptotically stable if the following conditions are satisfied:

(a) $f(x, t)$ is continuously differentiable throughout the entire motion space with respect to the x_i;

(b) $f(0, t) \equiv 0$;

(c) there is a real symmetric constant matrix B such that the eigenvalues (depending on x_0 and t) of the matrix

$$M = \tfrac{1}{2}(J^T B + BJ) \qquad (0 \leq t_0 \leq t < \infty) \tag{11.4}$$

lie below a fixed negative bound $-\delta$ for all x of a certain domain $\mathfrak{K}_r$; in (11.4), J is the functional matrix defined above.

If the eigenvalues of M are smaller than $-\delta$ for all x of the phase space, the equilibrium is asymptotically stable in the whole.

Proof: The following proof applies to autonomous differential equations. The real parts of the eigenvalues of the matrix BJ lie between the greatest and the smallest eigenvalue of its symmetric part M, i.e., they are smaller than $-\delta$. Hence in $\mathfrak{K}_r$

$$|\det BJ| \geq \delta^n$$

and in $\mathfrak{K}_r$ the function

$$|\det J| = \omega(r) \tag{11.5}$$

is always greater than a fixed positive number η. Therefore, the mapping of the x-space onto the f-space, determined by $f = f(x)$, is reversible in a certain neighborhood of the origin. This means that the point $x = 0$ is an isolated equilibrium of Eq. (11.3). Consequently, the Liapunov function

$$v = f^T(x)\, Bf(x) \tag{11.6}$$

is positive definite as a function of x. Its total derivative for (11.3)

$$\dot{v} = f^T(x)\, (J^T B + BJ)\, f(x)$$

is negative definite according to the assumption on M. But so far, this proves the asymptotic stability only "in the small." If $\omega(x) > \eta$ holds everywhere, we see that equation (11.3) has no other equilibrium than $x = 0$ in every finite subspace of the phase space. Furthermore, we obtain by integrating over the volume element in the f-space and in the x-space,

$$\int df = \int \omega(x)\, dx \geq \eta \int dx$$

From this it follows that at least one of the components of f increases beyond any bound as $|x|$ increases. Hence the function (11.6) is radially unbounded, and, as a consequence of Theorem 4.3, the equilibrium of (11.3) is asymptotically stable in the whole. The extension of this proof for the nonautonomous case is given by Krasovskii [21]; in this proof the lemma mentioned in Sec. 4, 8 is used (for a simpler proof cf. Ingwerson [1]).

In Theorem 11.1, instead of matrix (11.4) we may use a matrix N defined as follows. We determine the matrix $A(x, t)$ by the equation

$$AJ + J^T A = -U$$

and we put

$$C = \sum_{j=1}^{n} (\partial A/\partial x_j) f_j(x, t)$$

Then let

$$N = C + (\partial A/\partial t) - U$$

(Krasovskii [21]).

For the following nonautonomous differential equation, written in matrix form

$$\dot{x} = F(x, t) x \tag{11.7}$$

Zubov [1] proved the following:

THEOREM 11.2: The equilibrium of (11.7) is stable if all eigenvalues (depending on x and t) of the matrix

$$\tfrac{1}{2}(F^T + F)$$

are nonpositive in a certain domain $\Re_{h,t_0}$. The equilibrium is asymptotically stable if the eigenvalues are smaller than a quantity $-\delta(t)$, and if the integral

$$\int_{t_0}^{\infty} \delta(t)\, dt$$

diverges. The equilibrium is also asymptotically stable if $\lim F(0, t)$ exists as $t \to \infty$ and if all eigenvalues corresponding to this limit have negative real parts.

The first part of the theorem can be proved by means of the Liapunov function $|x|^2$, by means of Theorems 4.1, 4.2, and Remark 4, Sec. 4. The proof of the second part can be accomplished by a reasoning similar to that of Sec. 10(c).

The proof of Theorem 11.1 can easily be modified so as to obtain an estimate for the domain of attraction (Krasovskii [7] for $n = 2$, [10] for the general case). Let (11.3) be autonomous, and let the matrix B defined in Theorem 11.1 be equal to the unit matrix, U. Furthermore, let

$$m(r) = \min |f|, \qquad \mu(r) = \max |f|$$

and let $\lambda(r)$ be the minimum of the absolute value of the eigenvalues of (11.4) for $|x| = r$. Then the following theorem holds.

THEOREM 11.3 (Krasovskii [7, 10]): The spherical domain $|x| \leq r$ lies completely in the interior of the domain of attraction of the origin of (11.3) if the eigenvalues of (11.4) are negative in every domain $\Re_h$ for which the inequality

$$\int_r^h m(s)\,\lambda(s)\,ds > \tfrac{1}{2}\mu(r) \tag{11.8}$$

holds. If this integral diverges as $r \to 0$, $h \to \infty$, and if the eigenvalues of (11.4) are negative in the entire phase space, then the equilibrium is asymptotically stable in the whole.

The proof can be given by means of the Liapunov function $|f|^2$. The integral condition guarantees that the domain $\Re_h$ is not contained in the domain of attraction of any singularity of (11.3) which is different from 0. Another theorem by Krasovskii [7, 20] permits the condition of differentiability of Theorem 11.1 to be replaced by a somewhat weaker condition.

Estimates of the stability domain were also investigated in papers by Zubov (to be discussed in Sec. 21), Nemyckii [2] and Kim [1]. In the case where A in (10.2) is a triangular matrix, a better result can sometimes be obtained by means of a theorem of Perron [1]. A simple technical application is given by Vorel [1].

Recently, Ingwerson [1] extended the method of Sec. 8 to nonlinear systems, starting from the functional matrix $J(x)$. In a similar way, Schultz and Gibson [1] tackled the problem of generating Liapunov functions.

12. ESTIMATES FOR THE STABILITY DOMAIN OF THE PARAMETERS

Besides statements about stability of the equilibrium, the methods of Sec. 10, especially 10(b), and 10(c), furnish a tool to estimate the stability domain of the parameters which was defined in Sec. 3. The parameters may be functions of t. Equations treated by the direct method and the corresponding results are compiled in the sequel. Most cases deal with practical problems of nonlinear mechanics. Therefore, the equations are often written as scalar equations of higher order, rather than in vector form.

(a) Starzinskii [1] chose a time independent quadratic form in x and $\dot{x}$ as a Liapunov function to investigate the equation

$$\ddot{x} + p(t)\,\dot{x} + q(t)\,x = 0 \tag{12.1}$$

where $p(t)$ and $q(t)$ are continuous for $t > 0$ and satisfy the inequalities

$$0 \leq l_1 \leq p(t) \leq l_2, \qquad 0 \leq m_1 \leq q(t) \leq m_2$$

Evaluating the relations among the coefficients obtained for $\dot{v} = 0$, he obtained the inequalities

$$l_2 < \frac{m_2 + 2\sqrt{m_1 m_2} + 5m_1}{\sqrt{m_2} - \sqrt{m_1}}, \qquad l_1 > \sqrt{m_2} - \sqrt{m_1} \tag{12.2}$$

as sufficient conditions for the asymptotic stability of the equilibrium. The "Mathieu equation with a damping term"

$$\ddot{x} + a\dot{x} + \left(1 + b \cos \frac{2}{\mu} t\right) x = 0 \qquad (12.3)$$

(a and b constants, $\quad 0 < a, 0 < b < 1$)

is a special case of (12.1). This case has also been treated by Aizerman [3]; however, the condition

$$a > \sqrt{1+b} - \sqrt{1-b}$$

obtained from the relations (12.2), furnishes a better result than the Aizerman condition. The time dependent Liapunov function

$$v = x^2 + \frac{1}{q(t)} \dot{x}^2$$

results in the bound

$$a > \frac{b}{\mu \sqrt{1-b^2}}$$

which is in some cases even better.

(b) Starzinskii [2, 3] also dealt with equations of third or fourth order; however, the computations became rather complicated. A sufficient condition for asymptotic stability of the equilibrium of the particular differential equation

$$x^{(3)} + p\ddot{x} + \dot{x} + r(t) x = 0 \qquad (0 < p, 0 \leq r(t) \leq \rho)$$

is given as

$$\rho \leq p - (p^3/4), \qquad p \leq \sqrt{2}$$

or

$$\rho \leq 1/p, \qquad p \geq \sqrt{2}$$

Sufficient conditions for the differential equation

$$x^{(4)} + px^{(3)} + q\ddot{x} + \dot{x} + s(t) = 0 \qquad (0 < p, 0 < q, 0 \leq s(t) \leq \sigma)$$

are

$$\left(\frac{p^3}{4} - pq + 2\right) p \leq \sigma \leq \frac{1}{q}$$

$$\text{for} \quad \frac{1}{8p} (p^3 + 8 + \sqrt{p^6 + 16p^3}) \leq q \leq \frac{p^2}{4} + \frac{2}{p}$$

or

$$\sigma \leq 1/q \quad \text{for} \quad q \geq (p^2/4) + (2/p)$$

(c) Occasionally, parameter dependent, nonautonomous, linear equations occur which become independent of t for certain values of the parameter. For example,

$$\dot{x} = (A + \gamma P(t))\, x \qquad (12.4)$$

is such an equation. Let $P(t)$ be bounded and let γ be the parameter with $\gamma = 0$ as the special value for which (12.4) becomes independent of t. We compare (12.4) with (8.1) and use the Liapunov function (8.2). Its total derivative for (12.4) is

$$\dot{v} = -x^T(C - \gamma(P^T B + BP))\, x$$

The condition that $\dot{v}$ be negative definite for $t \geq t_0$ results in bounds for the parameter γ (Chetaev [8]).

G. A. Riabov [1] treated a special case. He considered a so-called "rectilinear particular solution" to the three-body problem. He established the differential equation of the perturbed motion, which is equivalent to the system

$$\dot{x}_1 = x_2, \qquad \dot{x}_2 = 2x_4 + \frac{\alpha}{1 + \gamma \cos t}\, x_1$$

$$\dot{x}_3 = x_4, \qquad \dot{x}_4 = -2x_2 - \frac{\beta}{1 + \gamma \cos t}\, x_3$$

Here, α, β, and γ are certain constants; t denotes the angle formed by the rectilinear motion and a direction determined by the initial values of the problem. The question of interest is, which values of γ result in unstable motions. Riabov answered this question by means of the method sketched above and by applying Theorem 5.1.

(d) Razumichin [8] considered a linear differential equation

$$\dot{x} = A(t)\, x \qquad (12.5)$$

and interpreted the matrix $A(t) = (a_{ik}(t))$ for $t \geq t_0$ as a parameter representation of a curve in the n^2-dimensional space of the coefficients a_{ik}. Let $v(x, t)$ be a positive definite function, and let $\dot{v}$ be its total derivative for (12.5). If this derivative is nonpositive for the value $t_1 \geq t_0$, then the corresponding "point" $A(t_1)$ shall belong *to the domain* $\mathfrak{L}(v)$. Thus, the domain $\mathfrak{L}(v)$ is the set of all points of the space of the coefficients where $\dot{v} \leq 0$. Evidently, a sufficient condition for the stability of the equilibrium of (12.5) is the existence of a definite function v, whose domain $\mathfrak{L}(v)$ contains the curve $A(t)$ ($t \geq t_0$) completely. If such a function is known, and if the corresponding domain $\mathfrak{L}(v)$ can be described, or at least estimated, then a sufficient stability condition is obtained in the form of inequalities for the $a_{ik}(t)$. For a quadratic form $v = x^T Bx$ (B constant), for example, $\mathfrak{L}(v)$ is of the type of an elliptic paraboloid. By this method, Razumichin [8] derived stability conditions for the differential equation

$$x^{(n)} + a_1 x^{(n-1)} + \cdots + a_{n-1}\dot{x} + a_n(t)\, x = 0$$

where only the lowest coefficient $a_n(t)$ is not constant.

13. THE PROBLEM OF AIZERMAN AND ITS MODIFICATIONS

(a) The problem of Aizerman belongs to the class of problems posed under Remark 4 of Sec. 6 and has initiated a large number of specific investigations. The starting point is the linear auxiliary equation (10.5) which was used in the discussion of Eq. (10.2). Estimates for the elements of the matrix G were obtained by means of the Liapunov function (8.2). These estimates also resulted in conditions for the nonlinear part g(x). It is evident that the indicated procedure does not yield the best possible estimates for the elements of G (cf. Sec. 10, last part of (b)). On the other hand, exact bounds for the elements of G can be obtained by utilizing the linearity of (10.5) and by applying Theorem 8.1. Investigation of the Hurwitz inequalities for the characteristic equation of the matrix A + G (cf. Malkin [19]) yields the exact stability domain of Eq. (10.5) in the g_{ik}-space, and, in the case of a *linear* function g(x), the exact stability domain for (10.2). The question now arises as to whether these exact bounds can also be applied to *nonlinear* additional terms; in other words, whether the bounds which are admissible for the elements g'_{ik}, g''_{ik} in the inequalities (10.4) coincide with the exact bounds for the additional linear terms.

This question is the content of Aizerman's problem in its most general form. The original formulation by Aizerman [1] was given in much more specific terms; namely, under the condition that the vector g(x) possess only a single nonvanishing component depending on a single argument only. In this particular case, the problem can be stated as follows:

Given the scalar differential equations

$$\dot{x}_1 = \sum_{k=1}^{n} a_{1k}x_k + f_1(x_j) \qquad (f_1(0) = 0, f_1(x) \in E)$$

$$\dot{x}_i = \sum_{k=1}^{n} a_{ik}x_k \qquad (i = 2, \ldots, n) \qquad (13.1)$$

where j denotes a fixed number between 1 and n. In addition, a linear comparison system is considered where the function $f_1(x_j)$ is replaced by ax_j. The inequality

$$\alpha x_j^2 < x_j f_1(x_j) < \beta x_j^2 \qquad (x_j \neq 0) \qquad (13.2)$$

shall be valid for every pair of numbers α, β for which asymptotic stability of the equilibrium of the linear comparison system follows from

$$\alpha < a < \beta \qquad (13.3)$$

Can it be inferred under these assumptions that the equilibrium of (13.1) is asymptotically stable in the whole? (As already mentioned, the maximum interval for admissible values of a follows from the Hurwitz conditions.)

The Aizerman problem stems from a question posed for a concrete system. The differential equations (13.1) can be considered as the equations of motion of an automatic control system with a single nonlinear element, as frequently occurs in practice. Here, asymptotic stability in the whole means that the control action decays after any arbitrary, finite, initial displacements have been imposed. Inequality (13.2) indicates that the only condition imposed on the characteristic of the nonlinear element is that it is to remain in a fixed angular domain.

So far, the problem has been completely solved only for the case of $n = 2$, and in that instance it was found to have an affirmative answer, provided an exceptional case of no practical interest is disregarded. Originally, it was conjectured that the general problem would also have an affirmative answer; however, the following compilation of results shows that additional conditions must be imposed on the nonlinearity even for the case $n = 3$. The same is true to an even greater extent if several equations are nonlinear and if the nonlinear functions depend on several variables.

(b) The first complete results were given by Erugin [2] and Malkin [16] and refer to the equations

$$\dot{x} = f(x) + by \quad (f(0) = 0)$$
$$\dot{y} = cx + dy \tag{13.4}$$

If $f(x)$ is replaced by hx, where h is a constant, the Hurwitz conditions for the linear comparison system are

$$h + d < 0, \quad hd - bc > 0 \tag{13.5}$$

The condition on the nonlinearity corresponding to (13.2) is

$$x(f(x) + dx) < 0, \quad x(df(x) - bcx) > 0 \quad (x \neq 0) \tag{13.6}$$

Providing $d^2 + bc \neq 0$, these inequalities are actually sufficient for the equilibrium of (13.4) to be asymptotically stable in the whole. This result is valid even if $f(x)$ is continuous only, i.e., if the question of the uniqueness of the solution is neglected (Erugin [5], Skackov [1]).

According to Malkin [16], the proof is accomplished by use of the positive definite Liapunov function

$$v = \int_0^x (f(\xi) d - bc\xi) \, d\xi + \tfrac{1}{2}(dx - by)^2 \tag{13.7}$$

whose derivative for (13.4),

$$\dot{v} = (f(x) + dx)(df(x) - bcx)$$

is nonpositive as a consequence of (13.6). Stability of the equilibrium follows immediately, and, according to Remark 7 of Sec. 4, asymptotic stability follows if $f(x) \in C_1$. To assure asymptotic stability in the whole (which is then also valid for the exceptional case $d^2 + bc = 0$) one has to require in addition that the expression

$$\gamma = \lim \int_0^x (f(\xi)\,d - bc\xi)\,d\xi \qquad (|x| \to \infty) \tag{13.8}$$

be unbounded. v is then radially unbounded. This assumption is satisfied, for example, if the inequality

$$d\frac{f(x)}{x} - bc \geq \delta > 0 \qquad (\delta \text{ fixed})$$

is valid for all large $|x|$.

The discussion of the Liapunov function (13.7) does not furnish the entire statement formulated in connection with (13.6). To obtain the entire statement, the phase curves must be studied in greater detail, and, as a consequence, resort must be made to qualitative methods (cf. the Introduction). Erugin [5] obtained the following supplementary results by such methods: If the quantity γ, defined by (13.8), is finite for the exceptional case $d^2 + bc = 0$ (i.e., if the equilibrium is not asymptotically stable in the whole), then the domain of attraction of the origin includes the domain $v < \gamma$. Pliss [1] showed that this domain is generally unbounded; but it is a proper subdomain of the phase plane. He also provided a procedure enabling the domain of attraction to be constructed with any desired accuracy. Krasovskii [1] gave a counter example illustrating that, in the exceptional case, the inequalities (13.6) are not sufficient for asymptotic stability in the whole. Let

$$\dot{x} = f(x) + y, \qquad \dot{y} = -x - y$$

with

$$f(x) = x - \frac{e^{-2x}}{1 + e^{-x}} \quad (x \geq 1), \qquad f(x) = x - \frac{e^{-2}}{1 + e^{-1}}x \quad (x < 1)$$

Apparently, $d^2 + bc = 0$ and $f(x) \in C_0$. The phase curve originating from the point $(1; e^{-1} - 1)$ has the equation $y = e^{-x} - x$; it does not approach the origin with increasing t.

(c) The second case of Aizerman's problem for $n = 2$ concerns the system

$$\dot{x} = ax + f(y), \qquad \dot{y} = cx + dy \tag{13.9}$$

Malkin [16] also investigated this system. It is a subcase of a more general system with two nonlinear functions, studied by Krasovskii [1, 3, 4]. The procedure is similar to that of (b) with the direct method being used first and the results then extended by qualitative discussions.

The most important results of Krasovskii are given below. It is assumed that the nonlinear functions $f_i(z) = zh_i(z)$ belong to the class C_1 and that these functions vanish at the origin. The following statements on the functions h_i shall always refer to nonvanishing arguments.

1. Let
$$\dot{x} = f_1(x) + by, \qquad \dot{y} = f_2(x) + dy$$
and
$$dh_1(x) - bh_2(x) > 0, \qquad h_1(x) + d < 0$$

Then the condition
$$\lim \left[(f_1(x) + dx) \operatorname{sgn} x - \int_0^x (f_1(\xi) d - f_2(\xi) b) d\xi \right] = -\infty \qquad (|x| \to \infty)$$

is necessary and sufficient for asymptotic stability in the whole. The function $\operatorname{sgn} x$ has the value of $+1$ for positive x, and -1 for negative x. Its definition for $x = 0$ is here and in the following of no importance.

In the proof, Krasovskii [1] used the Liapunov function
$$v = \int_0^x (f_1(\xi) d - f_2(\xi) b) d\xi + \tfrac{1}{2}(dx - by)^2$$
for which
$$\dot{v} = x^2 (h_1(x) + d)(dh_1(x) - bh_2(x))$$

In the linear case, the inequalities in the assumption reduce to the Hurwitz inequalities. If only one of the functions $f_1(x)$, or $f_2(x)$ is nonlinear, then the problem is equivalent to that of Eq. (13.4), or of Eq. (13.9), i.e., to the Aizerman problem for $n = 2$.

Krasovskii [3] proved the necessity of the conditions by employing qualitative methods.

2. Let
$$\dot{x} = f_1(x) + by, \qquad \dot{y} = cx + f_2(y)$$
be two nonlinear differential equations. Then the two conditions
$$h_1(x) + h_2(y) < 0, \qquad h_1(x) h_2(y) - bc > 0$$
are sufficient for asymptotic stability in the whole.

The proof is based on the Liapunov function
$$\tfrac{1}{2}(\beta^2 - bc) x^2 + \tfrac{1}{2}\left(b^2 - \frac{b^3 c}{\alpha^2}\right)^2 + \beta \int_0^x f_1(\xi) d\xi + \frac{b^2}{\alpha} \int_0^y f_2(\eta) d\eta - b\beta xy$$

depending on two parameters α and β chosen such that the inequalities
$$h_1(x) + \beta \leq 0, \qquad \beta h_1(x) - bc \geq 0$$
$$h_2(y) + \alpha \leq 0, \qquad \alpha h_2(y) - bc \geq 0$$
hold.

3. Let
$$\dot{x} = ax + f_2(y), \qquad \dot{y} = f_1(x) + dy$$
Sufficient conditions for asymptotic stability in the whole are:

(a) for $a < 0$, $d < 0$ the inequalities
$$h_1(x) \leq 0, \qquad h_2(y) \geq 0$$

(b) for $ad < 0$ the inequalities
$$a + d < 0, \qquad ad - h_1(x) h_2(y) > 0$$

4. Let
$$\dot{x} = f_1(x) + f_2(y), \qquad \dot{y} = cx + dy$$

The function $f_1(x) + dx$ be monotonically decreasing, and
$$dh_1(x) - ch_2(y) > 0$$

Then, if $d > 0$ and $c < 0$, the equilibrium is asymptotically stable in the whole. The case $c > 0$, $d < 0$ requires the additional condition

$$\lim \left| \int_0^y (f_2(\eta) - \beta\eta)\, d\eta \right| = \infty \qquad (|y| \to \infty)$$

where β is the parameter introduced under 2. Here, the conditions corresponding to the Hurwitz inequalities,

$$h_1(x) + d \leqq -\delta < 0, \qquad h_1(x) d - h_2(y) c \geqq \delta' > 0$$

are in this stronger form not even sufficient for asymptotic stability in the small.

Krasovskii [3] also solved some additional cases of the systems discussed under 2 to 4 not included in the previous compilation.

In this connection, Mufti [1] may also be mentioned, whose results were obviously found independently of Krasovskii.

(d) Tuzov [1] studied an "Aizerman system" of three differential equations, where the first equation is of the form

$$\dot{x} = a_{11}x + a_{12}y + a_{13}z + f(x) \qquad (f(0) = 0, f \in C_1)$$

The other two equations are assumed to be linear. Depending on the inequalities between the coefficients a_{ij} and the Hurwitz determinants, 22 subcases have to be distinguished. Tuzov constructed a Liapunov function for each case and arrived at stability conditions of the type discussed under (c).

Some other special systems were investigated by Tuzov [2] and Pliss [2, 3, 4]. Reference is also made to the summarizing monograph of Pliss [5] on Aizerman's problem for systems of two or three differential equations.

The Aizerman problem has not yet been completely solved even for the case $n = 3$. In principle, the methods developed so far can also be applied to systems of higher order; however, the rapidly increasing number of subcases would render the calculations so cumbersome and tedious that this approach to solving the general problem seems to promise little success. A review of the progress achieved until 1954 is given in a report by Erugin [7], (cf. also Nemyckii [1], Pliss [5]).

(e) As shown by the papers of Krasovskii [1, 3], the methods employed in the study of the restricted Aizerman problem can also be applied to some more general differential equations involving several nonlinear functions, but the cases dealt with as yet, are relatively specific cases sometimes associated with concrete problems.

Krasovskii [2] considered a system of three equations:

$$\dot{x} = f_1(x) + a_{12}y + a_{13}z$$
$$\dot{y} = f_2(x) + a_{22}y + a_{23}z \quad (f_i(0) = 0, f_i(x) \in E)$$
$$\dot{z} = f_3(x) + a_{32}y + a_{33}z$$

where $f_i(x) = xh_i(x)$. The cofactors belonging to the elements of the matrix

$$\mathsf{M}(x) = \begin{pmatrix} h_1 & a_{12} & a_{13} \\ h_2 & a_{22} & a_{23} \\ h_3 & a_{32} & a_{33} \end{pmatrix}$$

shall be denoted by d_{ik}. The characteristic equation of M is written in the form

$$\det(\mathsf{M} - \lambda \mathsf{U}) = -(\lambda^3 + a(x)\lambda^2 + b(x)\lambda + c(x)) = 0$$

The coefficients of this equation shall satisfy the "Hurwitz inequalities"

$$a(x) > 0, \quad a(x)b(x) - c(x) > 0, \quad c(x) > 0$$

for all $x \neq 0$. Then the following two theorems are valid:

(a) Let

$$a_{12}/a_{13} = d_{21}/d_{31} \qquad (13.10)$$

The condition

$$\lim \left[x \left(\frac{d_{21}}{a_{12}} - a(x) \right) \operatorname{sgn} x - \int_0^x \xi c(\xi)\, d\xi \right] = -\infty \qquad (|x| \to \infty)$$

is necessary and sufficient for asymptotic stability in the whole.

(b) Suppose, (13.10) is not valid. Then the two conditions

$$\lim \int_0^x \xi c(\xi)\, d\xi = \infty \qquad (|x| \to \infty)$$

$$4k_2 c(x)[a(x) - k_1] - \left[\frac{c(x)(1 + k_2)}{k_1} + k_2(a(x) - k_1)k_1 - k_2 b(x) \right]^2 > 0$$

are sufficient for asymptotic stability in the whole. k_1 and k_2 are two arbitrary, but fixed, positive numbers.

The proof is accomplished by the use of the direct method in conjunction with qualitative methods. The system is first transformed into a more convenient form by means of linear transformations. The distinction between the different cases is a consequence of these transformations.

Barbašin [2] considered the differential equation

$$x^{(3)} + a\ddot{x} + \varphi(\dot{x}) + f(x) = 0$$

in which $f(x)$ is continuously differentiable, $\varphi(y)$ is continuous, and both functions vanish if their arguments are zero. Sufficient conditions for asymptotic stability in the whole are obtained by means of the Liapunov function

$$v(x, y, z) = a \int_0^x f(\xi) \, d\xi + y f(x) + \int_0^y \varphi(\eta) \, d\eta + z^2/2$$

These conditions are:

(α) $a > 0$, $\dfrac{f(x)}{x} > 0 \; (x \neq 0)$, $a \dfrac{\varphi(y)}{y} - f'(x) > 0 \; (y \neq 0)$,

(β) $v(x, y, 0)$ radially unbounded in the (x, y)-plane.

They are satisfied, for instance, if two positive numbers h_1 and h_2 exist such that

$$\frac{f(x)}{x} \geq h_1 > 0, \qquad a \frac{\varphi(y)}{y} - f'(x) \geq a h_2 - h_1 > 0$$

M. L. Cartwright [1] provided some analogous sufficient conditions for the equations

$$x^{(4)} + a_1 x^{(3)} + a_2 \ddot{x} + a_3 \dot{x} + f(x) = 0$$

and

$$x^{(4)} + a_1 x^{(3)} + \varphi(\dot{x}) \ddot{x} + a_3 \dot{x} + a_4 x = 0$$

Here, it is assumed that $f(x) \in C_2$ and $\varphi(y) \in C_1$. Ogurtsov [1] investigated similar equations.

Nonlinear functions of several variables occur in papers of Šimanov [1, 2]. In the first paper, he considered the differential equation

$$x^{(3)} + f(x, \dot{x}) \ddot{x} + b \dot{x} + c x = 0$$

under the supposition that $f(x, y)$ and $f_x(x, y)$ are continuous. He arrived at the inequalities

$$b > 0, \quad c > 0, \quad f(x, y) > c/b, \quad y f_x(x, y) \leq 0$$

as sufficient conditions for asymptotic stability in the whole.

In the second paper he treated the system of differential equations

$$\ddot{x} + f_1(x, y, \dot{x}, \dot{y}) \dot{x} + \varphi_1(x) = 0 \qquad (\varphi_1(0) = 0)$$
$$\ddot{y} + f_2(x, y, \dot{x}, \dot{y}) \dot{y} + \varphi_2(y) = 0 \qquad (\varphi_2(0) = 0)$$

and obtained

$$\varphi_i(z)/z > 0 \; (z \neq 0), \quad f_i(x, y, \dot{x}, \dot{y}) > 0,$$
$$\int_0^z \varphi_i(\xi) \, d\xi \to \infty \; (|z| \to \infty) \qquad (i = 1, 2)$$

as sufficient conditions for asymptotic stability in the whole. Similar problems were dealt with by Ezeilo [1].

Razumichin [1] investigated the differential equation

$$\ddot{x} + \varphi(x, \dot{x}) \dot{x} + f(x, \dot{x}) = 0$$

but obtained complete results only in certain special cases.

14. THE PROBLEM OF LUR'E AND ITS GENERALIZATIONS

(a) The second method finds an important application in a problem belonging to the theory of automatic control. The problem probably was first formulated by Lur'e and Postnikov [1] (cf. also Lur'e [1]), and subsequently became the subject of numerous publications by, primarily, Soviet authors. Considered from a mathematical standpoint, we are concerned with the following problem.

Let the equations of motion for the quantities $x_1, \ldots, x_n, y, s$ be of the form

$$\dot{x} = \mathsf{A}x + \mathsf{p}y, \quad \dot{y} = f(s), \quad s = \mathsf{b}^T x = \sum_{i=1}^{n} b_i x_i \qquad (14.1)$$

where A is a constant matrix, $\mathsf{p} \neq 0$ is a constant vector, and $f(s)$ is a nonlinear function with the properties

$$f(s) \in E, \quad f(0) = 0, \quad sf(s) > 0 \quad \text{for} \quad s \neq 0 \qquad (14.2)$$

The components $b_1, \ldots, b_n$ of b shall be determined such that the equilibrium of Eq. (14.1) is *absolutely stable*, in accordance with the following definition:

DEFINITION 14.1: The equilibrium of (14.1) is called *absolutely stable* if it is asymptotically stable in the whole for all functions $f(s)$ possessing the properties (14.2).

System (14.1) describes the behavior of a control system without auxiliary energy; x comprises the general coordinates, y is the controller, $f(s)$ the nonlinear characteristic of the control motor of proportioned action and s is the input signal of the motor. It is desired to obtain the stability domain in the space of the parameters b_i of the controller for all functions $f(s)$ admissible in the sense of (14.2).

The more general system

$$\dot{x} = \mathsf{A}x + \mathsf{p}y, \quad \dot{y} = f(s), \quad s = \mathsf{b}^T x - hy \quad (h > 0) \qquad (14.3)$$

can be written in place of (14.1). It describes a control system with auxiliary energy and a control motor of integral action. This system can formally be transformed into the form (14.1) by substituting new variables. But the matrix of the linear principal part will thereby be transformed into a matrix with a vanishing eigenvalue. Obviously, this has an effect on solving the problem. Therefore, a separate treatment is recommended.

(b) The procedure of solving Lur'e's problem for the differential equations (14.3) is given below. A complete treatment can be found in papers of Lur'e [7] and Letov [8]. Summaries of the literature were given by Lur'e [8] and Letov [9, 11]; cf. also Yakubovich [2] and the very elegant treatment given by Lefschetz [3].

The differential equations are at first transformed into the so-called canonical form by means of a transformation developed by Lur'e [3]. For this purpose, the eigenvalues $\alpha_1, \ldots, \alpha_n$ of A are assumed to be simple. Let D be the diagonal matrix with the diagonal elements $\alpha_1, \ldots, \alpha_n$. Further, let n be the column vector with the elements $1, \ldots, 1$. If $\det A \neq 0$, a vector q and a matrix B can be determined by the conditions

$$B^I AB = D, \quad Aq + p = 0, \quad B^I q = -n$$

Then by setting

$$x = Bz + yq, \quad b^T AB = x^T$$

the Eqs. (14.3) are transformed into

$$\dot{z} = Dz + f(s) n, \quad \dot{s} = x^T z - hf(s) \qquad (14.4)$$

If $\det A = 0$, Eqs. (14.4) can be obtained by putting

$$z = Cx + yn, \quad b^T AC^I = x^T$$

where

$$CAC^I = D, \quad Cp = Dn$$

Lur'e [3, 7] and Letov [8] supplied explicit transformation formulae for (14.1) and (14.3). The transformation can still be accomplished if the eigenvalues of A are not all simple (Troickii [1], Spasskii [1, 2]). The matrix D is then replaced by the Jordan normal form of A.

The matrix B can be computed as indicated only if p is orthogonal to none of the eigenvectors of A. However, Theorem 14.1 remains true in any case. In the exceptional case (not considered by Lur'e) where p is orthogonal to $l \geq 1$ eigenvectors of A, let m be the vector of $n - l$ components 1 and l components 0. One proceeds now with m in the same way as previously with n. If in the corresponding canonical form of the resulting equation l components of x are put equal to zero, the equation splits into two equations of the order $n - l$ and l, respectively, which are independent of each other. One of these equations is linear and it may be omitted from the next steps which can be carried out as in the normal case.

By eliminating at first y from the differential equations (14.3), these equations can be rewritten as

$$\dot{x} = (A + h^{-1}pb^T) x - h^{-1}sp$$
$$\dot{s} = b^T (A + h^{-1}pb^T) x - (s/h) b^T p - hf(s)$$

If the vector x is subjected to a linear transformation such that the matrix $A + h^{-1}pb^T$ is transformed into its Jordan normal form, then another suitable "canonical" form is obtained (Letov [4]).

In the sequel, the eigenvalues α_i shall be numbered so that the first k of them are real and the remaining are conjugate complex pairs:

$$\alpha_1, \ldots, \alpha_k \text{ real}, \quad \alpha_{k+1} = \bar{\alpha}_{k+2}, \ldots, \alpha_{n-1} = \bar{\alpha}_n \qquad (14.5)$$

As a consequence of (14.4), the same holds for the components of the vector z,

$$z_1, \ldots, z_k \text{ real}, \quad z_{k+1} = \bar{z}_{k+2}, \ldots, z_{n-1} = \bar{z}_n$$

Now, let

$$\operatorname{Re} \alpha_i < 0 \quad (i = 1, \ldots, n) \tag{14.6}$$

i.e., the equilibrium of the linear part (in other words, the equilibrium of the control loop with disconnected controller) is asymptotically stable. We introduce the vector g with the components $g_1, \ldots, g_n$ and the vector $w = w(t)$ with the components $e^{\alpha_i t} g_i$. For the time being, the g_i shall be considered as unknowns; but let

$$g_1, \ldots, g_k \text{ real}, \quad g_{k+1} = \bar{g}_{k+2}, \ldots, g_{n-1} = \bar{g}_n \tag{14.7}$$

Furthermore, let K be a diagonal matrix with positive, but still undetermined elements $k_1, \ldots, k_n$ and, finally, let

$$M = -(g_i g_k / (\alpha_i + \alpha_k))$$

By assumption (14.5), the quantity

$$(z^T w)^2 = z^T w w^T z$$

is real and nonnegative for finite values of t. The integral

$$\gamma(z) = \int_0^\infty (z^T w(t))^2 \, dt$$

converges with regard to (14.6) and it can be written as a quadratic form,

$$\gamma(z) = z^T M z$$

We construct the Liapunov function

$$v = \gamma(z) + \bar{z}^T K z + \int_0^s f(u) \, du \tag{14.8}$$

It follows from (14.2) and (14.6) that (14.8) is a positive definite function of the $n + 1$ real variables $z_1, \ldots, z_k$, Re z_{k+1}, Im z_{k+1}, Re $z_{k+3}, \ldots$, Im z_n, s. Further statements on the definiteness behavior shall refer to these variables. In order to get v radially unbounded, we additionally require

$$\int_0^\sigma f(u) \, du \to \infty \quad \text{as} \quad |u| \to \infty$$

The total derivative of v for (14.4) is

$$\dot{v} = z^T(DM + MD)z + \bar{z}^T(\bar{D}K + KD)z - hf^2 \\ + (n^T M z + z^T M n + n^T K z + \bar{z}^T K n + x^T z) f(s)$$

Now,

$$DM + MD = -gg^T, \quad \bar{z}^T(\bar{D}K + KD)z = 2\sum_{i=1}^n k_i \operatorname{Re} \alpha_i |z_i|^2$$

If the identically vanishing quantity

$$2f(s)\sqrt{\bar{h}}\,\mathsf{g}^T z - 2f(s)\sqrt{\bar{h}}\,\mathsf{g}^T z$$

is added to $\dot{v}$, then

$$\dot{v} = 2\sum_{i=1}^{n} k_i \operatorname{Re} \alpha_i |z_i|^2 - \left(f(s)\sqrt{\bar{h}} + \sum_{i=1}^{n} g_i z_i\right)^2$$

$$+ f(s)(\mathsf{x}^T z + 2\mathsf{n}^T \mathsf{M} z + 2\sqrt{\bar{h}}\,\mathsf{g}^T z + \mathsf{n}^T \mathsf{K}(z + \bar{z}))$$

As a consequence of the assumption (14.6), the first two terms are certainly negative. Therefore, the derivative $\dot{v}$ is negative definite, provided g is so determined that the factor of $f(s)$ vanishes identically. This leads to the determining equations.

$$r_i - 2g_i \sum_{j=1}^{n} \frac{g_j}{\alpha_i + \alpha_j} + 2\sqrt{\bar{h}}\, g_i + 2k_i = 0 \qquad (i = 1, \ldots, n) \quad (14.9)$$

If we put $k_i = 0 (i = 1, \ldots, n)$, all considerations remain valid; but $\dot{v}$ becomes negative semi-definite. Asymptotic stability in the whole can still be inferred by using Remark 7 of Sec. 4: $\dot{v}$ can be equal to zero only for $\sqrt{\bar{h}}\,f(s) + \mathsf{g}^T z = 0$, and this relation cannot be valid along a complete half-trajectory of (14.4), since otherwise a linear system of differential equations could be obtained by elimination of $f(s)$. This leads to:

THEOREM 14.1 (Lur'e): If the algebraic equations

$$r_i + 2\sqrt{\bar{h}}\, g_i - 2g_i \sum_{j=1}^{n} \frac{g_j}{\alpha_i + \alpha_j} = 0 \qquad (i = 1, \ldots, n) \quad (14.10)$$

can be solved with respect to the quantities $g_1, \ldots, g_n$ such that the solutions satisfy the conditions (14.7), then the equilibrium of (14.3) is absolutely stable.

The stability problem has thus been reduced to an algebraic problem.

(c) Multiplying the ith equation of (14.10) by α_i, or α_i^{-1}, respectively, and adding the resulting expressions, we obtain the relations

$$2\sqrt{\bar{h}} \sum_{i=1}^{n} g_i \alpha_i + \sum_{i=1}^{n} r_i \alpha_i - \left(\sum_{i=1}^{n} g_i\right)^2 = 0$$

$$2\sqrt{\bar{h}} \sum_{i=1}^{n} \frac{g_i}{\alpha_i} + \sum_{i=1}^{n} \frac{r_i}{\alpha_i} - \left(\sum_{i=1}^{n} \frac{g_i}{\alpha_i}\right)^2 \equiv \sum_{i=1}^{n} \frac{r_i}{\alpha_i} - \left(\sum_{i=1}^{n} \frac{g_i}{\alpha_i} - \sqrt{\bar{h}}\right)^2 + h = 0$$

It follows from the last relation that

$$h + \sum_{i=1}^{n} \frac{r_i}{\alpha_i} \geq 0$$

is a necessary condition for the solvability of (14.10). (Bromberg [1] made a critical remark regarding the sign of equality.)

A slight modification of (14.8) leads to results in the case of one vanishing eigenvalue, say, $\alpha_1 = 0$. Then $\gamma(z)$ is formed only for $z_2, \ldots, z_n$ with the result that the equations (14.9) and (14.10) are also obtained only for $i = 2, \ldots, n$. The relation

$$k_1 + r_1 = 0, \quad \text{i.e.,} \quad r_1 < 0$$

is obtained in place of Eq. (14.9). Inclusion of several eigenvalues with nonpositive real parts also causes no fundamental difficulties (Letov [3], Troickii [1]). By means of an algorithm given by Lur'e [5, 7], system (14.10) can be transformed into an equivalent system in which the coefficients of the original equation (14.3) occur instead of the elements of the transformed quantities D and x. Thereby, computation of the eigenvalues of A is avoided.

In the case $n = 2$, Lur'e's equations have the form

$$\frac{g_i^2}{\alpha_i} - 2\sqrt{h}\, g_i + \frac{2g_1 g_2}{\alpha_1 + \alpha_2} - r_i = 0 \quad (i = 1, 2)$$

If new variables t_1, t_2 are introduced by setting

$$g_i = -\alpha_i(t_i - \sqrt{h}) \quad (i = 1, 2)$$

the two equations

$$(t_1 + t_2 - \sqrt{h})^2 = \kappa^2$$

$$(\alpha_1 - \alpha_2) t_1^2 + 2\alpha_2(\sqrt{h} \pm \kappa) t_1 - \alpha_2(\sqrt{h} \pm \kappa)^2 - \alpha_1 + \alpha_2 - r_1 + r_2 = 0$$

are obtained, where the quantity

$$\kappa^2 = h + (r_1/\alpha_1) + (r_2/\alpha_2)$$

is obviously positive. If the side conditions (14.7) for the g_i and the conditions which follow from them for the t_i are considered, the desired sufficient conditions are obtained in the form

$$\alpha_1^2 + \alpha_2^2 + \alpha_1 r_1 + \alpha_2 r_2 \pm 2\alpha_1 \alpha_2 \kappa > 0 \qquad (14.11)$$

They determine a domain in the (r_1, r_2)—plane in which absolute stability is assured. (If α_1 and α_2 are real, (14.11) is immediately evident. That the condition is also valid for $\alpha_1 = \bar{\alpha}_2$ can be recognized by establishing separate equations for the real and imaginary parts.)

V.–M. Popov [1] carried out a somewhat more precise investigation of the case $n = 2$; he also arrived at necessary conditions.

The discussion of (14.10) becomes considerably more cumbersome even if $n = 3$; it has been completed only in special cases for $n \geq 4$ (Lur'e [5, 7]). Rozenvasser [1] discussed in detail the stability conditions for (14.1) in the cases $n = 5$ and $n = 6$.

(d) Another sufficient condition for absolute stability can be derived from (14.8) in a different manner (Malkin [10]). By setting $K = 0$ and by writing

$$-\dot{v} = hf^2 + \left(\sum_{i=1}^{n} g_i z_i\right)^2 - 2f \sum_{i=1}^{n} z_i u_i \quad \left(2u_i = r_i - 2g_i \sum_{j=1}^{n} \frac{g_j}{\alpha_i + \alpha_j}\right)$$

the right hand side can be regarded as a quadratic form in the variables f, $z_1, \ldots, z_k$, $\operatorname{Re} z_{k+1}, \ldots, \operatorname{Im} z_n$. Positive definiteness of this form assures stability. Now, the form has the discriminant

$$\begin{vmatrix} g_1^2 & g_1 g_2 & \cdots & g_1 g_n & u_1 \\ g_1 g_2 & g_2^2 & \cdots & g_2 g_n & u_2 \\ \cdot & \cdot & \cdots & \cdot & \cdot \\ g_1 g_n & g_2 g_n & \cdots & g_n^2 & u_n \\ u_1 & u_2 & \cdots & u_n & h \end{vmatrix} \qquad (14.12)$$

whose first n-principal minors are evidently positive. Thus, the only remaining sufficient stability condition is that under the side condition (14.7) the numbers $g_1, \ldots, g_n$ can be chosen such that the determinant (14.12) is positive. Investigations of special cases have shown that Theorem 14.1 furnishes sometimes better sometimes worse estimates than the evaluation of the conditions just mentioned (cf. also Komarnitskaia [1]).

(e) If the assumption (14.6) is not satisfied, but if eigenvalues with positive real parts occur, the system is said to be *inherently unstable* (Letov [4], Lur'e [6]). The problem formulated in connection with (14.2) can now only be solved by imposing additional requirements on the nonlinear function $f(s)$ beyond those of (14.2). According to Letov [4], a suitable additional condition is

$$f(s) = cs + \varphi(s) \qquad (c > 0,\ s\varphi(s) > 0 \text{ for } s \neq 0) \qquad (14.13)$$

Under this assumption, the stability problem can be approached in the following manner: We replace the nonlinearity in (14.3) by cs and consider the linear system of equations obtained by elimination of s,

$$\dot{\mathbf{x}} = \mathbf{A}\mathbf{x} + \mathbf{p}y, \qquad \dot{y} = c\mathbf{b}^T\mathbf{x} - chy$$

We also consider the corresponding original nonlinear system

$$\dot{\mathbf{x}} = \mathbf{A}\mathbf{x} + \mathbf{p}y, \qquad \dot{y} = c\mathbf{b}^T\mathbf{x} - chy + \varphi(s) \qquad (s = \mathbf{b}^T\mathbf{x}) \qquad (14.14)$$

and we treat these equations as a system of the type (14.1) in the $n+1$ variables $x_1, \ldots, x_n, y$. The matrix of the linear principal part can be written as

$$\mathbf{A}_1 = \begin{pmatrix} a_{11} & a_{12} & \cdots & a_{1n} & p_1 \\ a_{21} & a_{22} & \cdots & a_{2n} & p_2 \\ \cdot & \cdot & \cdots & \cdot & \cdot \\ a_{n1} & a_{n2} & \cdots & a_{nn} & p_n \\ cb_1 & cb_2 & \cdots & cb_n & -ch \end{pmatrix}$$

We now require at first that the matrix $\mathbf{A}_1$ have only eigenvalues with negative real parts; this leads to $n+1$ inequalities for c. Then (14.14) is treated as in (a) to obtain a system of quadratic equations of the type (14.10) which must be solvable with side conditions such as (14.7). The conditions for solvability and the stated inequalities represent the desired sufficient conditions for absolute stability of the equilibrium. Another way of approaching the problem was discussed recently by Rekasius and Gibson [1].

(f) In several variations, the differential equations (14.1) and (14.3) occur in control theory. These different forms of the equations necessitate slight modifications of the methods described in (a) to (c). Letov [2] considered a system in which the equation for $\dot{x}$ contains a constant perturbation term, such that the "trivial solution" becomes a constant different from zero. In this case, the differential equations of the perturbed motion must first be established. Lur'e [2] investigated a system with the nonlinearity $f(s) = c(\text{sgn } s + \text{sgn } \dot{s})$ which is discontinuous at the origin. A Liapunov function of the type (14.8) is adequate for a system of this type; $\dot{v}$, naturally, becomes discontinuous. However, the system has equilibria other than the trivial solution and thereby necessitates a special consideration. Vorovich [1] assumed that (14.1) is subjected to certain random perturbations and he employed statistical methods to estimate their effect. Control systems with several controllers have been considered occasionally. The equations of motion analogous to (14.1) then become

$$\dot{x} = Ax + \sum_{j=1}^{m} f_j(s_j) p_j, \quad s_j = b_j^T x \quad (j = 1, \ldots, m) \quad (14.15)$$

This case does not introduce basically new concepts. The numerical evaluation of the stability conditions becomes quite cumbersome even for $m = 2$ (Duvakin and Letov [1], Letov [8], Spasskii [1], Rumiancev [3]).

If the feedback of the control loop contains a derivating term, then $\dot{y}$ occurs in the equation for s,

$$s = b^T x - hy - h_1 \dot{y} \quad (h \geq 0, h_1 \geq 0)$$

The term $h_1 \dot{y}$ has no effect on stability if all eigenvalues of A have negative real parts; however, it does become important for an inherently unstable system (Letov [10, 12], also Bass [1]). In a paper of Chang [1], the second equation in (14.3) is of the form $\ddot{y} + a\dot{y} + by = f(s)$. Lefschetz [3, 4] modified the original equations by adding more nonlinear terms (cf. also Yakubovich [3, 4]).

Razumichin [2] treated the differential equations (14.1) and (14.4) under a point of view which falls more into the realm of Aizerman's problem (Sec. 13). He put $f(s) = s\varphi(s)$ in (14.4) and he considered the system as linear in the variables z_i and s, with φ considered as a parameter. As a Liapunov function, a quadratic, or Hermitian form, respectively, is constructed for this system; one of its coefficients is treated as a variable parameter γ. To each value of γ there corresponds an interval $\varphi_{1\gamma} < \varphi < \varphi_{2\gamma}$ such that the linear system has an asymptotically stable equilibrium. That interval $[\varphi_1, \varphi_2]$ common to all intervals has the following property: If the inequality $\varphi_1 < \varphi(s) < \varphi_2$ is satisfied for all values of the argument s, then the equilibrium is asymptotically stable in the whole. Razumichin derived a quadratic equation for the computation of the numbers φ_1 and φ_2. If A has only real eigenvalues, the stability condition can be formulated even more conveniently. Skackov [2, 3] studied another modification of the problem. The system has the form

$$\dot{x}_i = a_i x_i + b_i y \quad (i = 1, 2) \qquad \dot{y} = f(x_1, x_2, y)$$

f vanishes on the plane $y = p_1 x_1 + p_2 x_2$.

$$1 + \left|\frac{b_1 p_1}{a_1}\right| + \left|\frac{b_2 p_2}{a_2}\right| > 0$$

is a sufficient condition for asymptotic stability in the whole. In another note, Skackov [4] showed that this inequality is necessary in a certain sense: If the left hand side is negative, functions f can be constructed such that the equilibrium of the system becomes unstable.

Lur'e's problem for nonautonomous differential equations shall be treated later in connection with a different subject (Sec. 34).

15. ESTIMATES FOR THE SOLUTIONS

Assume a Liapunov function v to be known for an autonomous differential equation with asymptotically stable equilibrium. Let $\dot{v}$ be negative definite in a certain neighborhood of the origin. From the qualitative discussion of Sec. 3 it follows that during the time interval $t_0 < t < t_1$, the trajectory $x(t)$ is located between the hyper-surfaces $v = v(x(t_0))$ and $v = v(x(t_1))$. If the distance of the hyper-surface $v = $ const from the origin can be computed or estimated, then estimates for $|x(t)|$ can be obtained. So far this possibility has not yet been systematically utilized; only several special results are known to date.

Chetaev [4, 8] considered Eq. (8.1) in the stable case by means of the Liapunov function v determined by (8.2) and put $C = U$ in (8.3). If λ_1 denotes the smallest, and λ_n the largest eigenvalue of the matrix B, then, as is well known,

$$\lambda_1 |x|^2 \leq v \leq \lambda_n |x|^2$$

Furthermore, according to the method of construction

$$\dot{v} = -|x|^2$$

Hence

$$v(x(0)) e^{-t/\lambda_1} \leq v(x(t)) \leq v(x(0)) e^{-t/\lambda_n}$$

APPLICATIONS OF THE STABILITY THEOREMS

$$|x(t)|^2 \leq \frac{v(x(t))}{\lambda_1} \leq \frac{v(x(0))}{\lambda_1} e^{-t/\lambda_n} \leq |x(0)|^2 \frac{\lambda_n}{\lambda_1} e^{-t/\lambda_n}$$

or finally,

$$|x(t)|^2 \leq |x(0)|^2 \frac{\lambda_n}{\lambda_1} e^{-t/\lambda_n}$$

For the motion to have the property that the image point in the phase space travels through the distance between the two spheres of radii

$$r_0 = |x(0)| \quad \text{and} \quad r_1 = |x(t)|$$

in a minimal time, the expression

$$\lambda_n \log \frac{r_0^2 \lambda_n}{r_1^2 \lambda_1}$$

must be as small as possible. Application of this rule requires a knowledge of the eigenvalues of the matrix B. Bedel'baev [1] provided a procedure which bypasses the direct computation of the eigenvalues.

Razumichin [5] treated the nonautonomous linear differential equation in a similar manner. Let $v = x^T B(t) x$ be the Liapunov function constructed in accordance with Sec. 9; the determinant of B shall be denoted by b, and the cofactor belonging to the element b_{kk} by β_k. Furthermore, let $\dot{v} = x^T H(t) x$, and let $\tilde{\mu}_0 = \tilde{\mu}_0(t)$ denote the largest root of the equation $\det(B - \mu H) = 0$. Then the inequality

$$|x_k(t)| \leq \sqrt{v(x_0) \frac{\beta_k}{b}} \exp\left(\tfrac{1}{2} \int_{t_0}^{t} \tilde{\mu}_0(\tau)\, d\tau\right)$$

is valid for the kth component x_k of the solution vector $p(t, x_0, t_0)$. From this inequality, incidentally, the stability condition (9.8) can be easily obtained. An equivalent estimate was given by Gorbunov [2, 4]. Letov [5] investigated the solution of a nonlinear equation of the type (14.3) with asymptotically stable equilibrium. He attempted to determine a number κ such that for $t_0 = 0$ the inequality

$$|x(t)| < |x_0| e^{-\kappa t}$$

always holds (cf. Sec. 25, concept of order number). Letov chose that instant τ for which

$$|x(\tau)| = e^{-1} |x_0|$$

as a measure of the rate of transient decay of the process described by $x(t)$ (in control engineering terms: as a measure of the quality of the control process).

Obviously,

$$\tau \leq 1/\kappa$$

Under the assumption that the nonlinearity is of the form (14.13), Letov first established a canonical form for the original equations (cf. Sec. 14 (b)):

$$\dot{z}_i = \beta_i z_i + s, \qquad \dot{s} = \sum_{i=1}^{n} r_i z_i - as - \varphi(s) \tag{15.1}$$

Introducing a parameter k by putting

$$y = e^{kt} z, \qquad \eta = e^{kt} s \tag{15.2}$$

the following system of differential equations for the variables $y_1, \ldots, y_n, \eta$ can be derived from (15.1):

$$\begin{aligned}\dot{y}_i &= (\beta_i + k) y_i + \eta \\ \dot{\eta} &= \sum_{i=1}^{n} r_i y_i - (a + k) \eta - e^{kt} \varphi(e^{-kt} \eta)\end{aligned} \tag{15.3}$$

For $k = 0$, the equilibrium of (15.3) is asymptotically stable by assumption, but for large values of k it is certainly unstable. Because of (15.2) the smallest value of k resulting in instability of the equilibrium of (15.3) is an upper bound for the desired number κ. The stability test can be carried out by means of the Liapunov function

$$v = |y|^2 + \eta^2$$

Similarly as for (14.12), several conditions follow from the inequality $\dot{v} < 0$. Only the last of these conditions is essential; it furnishes an algebraic equation of degree n whose smallest real root is an upper bound for κ. In special cases, this bound is the best which can be obtained.

Pliškin [1] investigated some differential equations, studied before by Krasovskii [1, 3] (cf. Sec. 13). He estimated the integrals

$$\int_0^\infty (x(t))^2 \, dt, \qquad \int_0^\infty ((x(t))^2 + (y(t))^2) \, dt \tag{15.4}$$

For example, under the assumption

$$dh_1(x) - bh_2(x) \geq \delta_1 > 0, \qquad h_1(x) + d \leq -\delta_2 < 0$$

one obtains for the system treated in Sec. 13 (c),1, the estimate

$$\dot{v} \leq -2\delta_1 \delta_2 x^2$$

from which it follows that

$$\int_0^\infty x^2 \, dt < \frac{v(x_0, y_0)}{2\delta_1 \delta_2}$$

Integrals of the type (15.4) play a role in interpreting the equations of motion under the point of view of control engineering. The numerical values of these integrals can be considered as a measure of the control quality. In the linear case, the relation

$$\int_0^\infty x^T(t) \mathsf{C} x(t) \, dt = x_0^T \mathsf{B} x_0 \tag{15.5}$$

is obtained by integrating (8.2) under consideration of (8.3). This is a closed expression for the "generalized quadratic control-surface" introduced by Feldbaum [1]. But the value of this expression for practical evaluation is somewhat problematic (cf. B. Herschel [1, 2]).

4 THE CONVERSE OF THE MAIN THEOREMS

16. STATEMENT OF THE PROBLEM

The main theorems on stability and instability introduced in Secs. 4 and 5 furnish only sufficient conditions; they contain no hints on how to find a Liapunov function for a given differential equation. In fact, the general question on the existence of such functions remains completely open. On the other hand, suitable functions can be determined for numerous specific cases and also for certain general types of differential equations (cf. Chap. 3). The question then arises as to what extent the main theorems can be reversed, i.e., under what circumstances the existence of a Liapunov function can be inferred assuming the stability behavior to be known. In other words, under what circumstances are the sufficient conditions expressed by the main theorems also necessary.

Only recently, this part of the theory of the direct method has received attention. Nevertheless, the main questions may be considered to be satisfactorily answered even though the theory has not yet been completely worked out. The first converse statement of the theorem on weak stability has been given by K. P. Persidskii [4]. Massera [1] achieved the first substantial success with respect to asymptotic stability, for which Malkin [1, 3] previously solved some specific cases. Massera proved the reversibility of Theorem 4.2 for autonomous and periodic differential equations. Malkin recognized that the most important point of Massera's proof is not the restriction to autonomous and periodic equations, but rather the concept of the "uniformity" of the stability (which is an a priori property of these types of equations (cf. Sec. 17). Malkin [20] proved a converse theorem for nonautonomous equations which has been improved in the meantime by Massera [4]. Other converse theorems have been given independently by Barbašin [1], Krasovskii [12, 16, 19], Kurzweil [1, 2, 3], and Zubov [6].

The investigations indicated that a refinement of the concepts "stable" and "asymptotically stable" is necessary to clarify the situation.

The point is whether or not the function

$$\mu(t, x_0, t_0) = \sup_{u \geq t} |p(t_0 + u, x_0, t_0)| \qquad (16.1)$$

can be uniformly estimated with respect to x_0 and, above all, with respect to t_0. The function (16.1) has the properties:

(a) in the case of stability, it is bounded with respect to t for fixed t_0 and x_0 and it tends monotonically towards 0 as $x \to 0$.

(b) in the case of quasi-asymptotic stability, it tends monotonically towards 0 as $t \to \infty$ (x_0 and t_0 being fixed). If the function (16.1) can be uniformly estimated, then the terminology *uniform stability*, and *uniform asymptotic stability* will be used, in contrast to *nonuniform* stability. It was found that the assumptions of Theorem 4.2 immediately guarantee uniform asymptotic stability. Theorem 4.2, as given in Sec. 4, is too weak; it cannot be expected to be reversible. Nevertheless, it can be reversed if it is expressed as a statement on uniform asymptotic stability.

Several methods of proving the converse theorems are now known. One of these methods shows the existence of a suitable Liapunov function by providing a construction procedure, essentially in an analytical manner. But the procedure is transfinite, i.e., "construction" cannot be understood as in Chap. 3. Another method uses topological tools, in particular the theory of dynamical systems (cf. Sec. 20). Finally, a method of Zubov [3, 6] has to be mentioned. Here, a Liapunov function is obtained by solving a partial differential equation.

The inversion of the theorems on instability has also been studied (Krasovskii [8], Vrkoč [1]). This problem is basically simpler, since instability of the equilibrium is a much less characteristic property than either stability, or asymptotic stability. A comparing survey of the converse theorems was given by Massera [6].

17. UNIFORM STABILITY

The number δ occuring in Definition 2.1 is in general not only a function of the given deviation ϵ, but it depends also on the initial instant t_0. In the case of *uniform stability*, the number δ depends only on ϵ and not on t_0.

DEFINITION 17.1 (K. P. Persidskii [2]): *The equilibrium of differential equation* (2.7) *is said to be uniformly stable if for each $\epsilon > 0$ a number $\delta = \delta(\epsilon) > 0$, depending only on ϵ, can be determined such that the inequality*

$$|p(t, x_0, t_0)| < \epsilon \qquad (t \geq t_0) \qquad (17.1)$$

follows from

$$|x_0| < \delta$$

for all $t_0 \geq 0$.

The definition can also be formulated by means of a *comparison function*, as can be seen from the following theorem.

THEOREM 17.1: *The equilibrium of differential equation* (2.7) *is uniformly stable if and only if there exists a function* $\rho(r)$ *with the following properties:*

(a) $\rho(r)$ *is defined, continuous, and monotonically increasing in an interval* $0 \leq r \leq r_1$;

(b) $\rho(0) = 0$; *the function* ρ, *therefore, belongs to the class* K;

(c) *the inequality*

$$|\mathsf{p}(t, \mathsf{x}_0, t_0)| \leq \rho(|\mathsf{x}_0|) \qquad (17.2)$$

is valid for $|\mathsf{x}_0| < r_1$.

It is immediately clear that the condition is sufficient. The necessity of the condition, i.e., the existence of a suitable function $\rho(r)$ in the case of uniform stability can be shown as follows: We consider for a given $\epsilon > 0$ the upper bound $\bar{\delta}(\epsilon)$ of all numbers δ belonging to ϵ due to Definition 17.1. If $|\mathsf{x}_0| \leq \bar{\delta}(\epsilon)$, then $|\mathsf{p}(t, \mathsf{x}_0, t_0)| \leq \epsilon$, and there exists at least one initial point $\tilde{\mathsf{x}}_0$ with $|\tilde{\mathsf{x}}_0| \leq \delta_1$ for every $\delta_1 > \bar{\delta}$ such that at some time t the quantity $|\mathsf{p}(t, \tilde{\mathsf{x}}_0, t_0)|$ exceeds the value ϵ. Obviously, the function $\bar{\delta}(\epsilon)$ is positive for $\epsilon > 0$; it is nondecreasing, it tends towards zero as $\epsilon \to 0$, and it is possibly discontinuous. We choose now a continuous, monotonically increasing function $\hat{\delta}(\epsilon)$ such that $\hat{\delta}(\epsilon) \leq \bar{\delta}(\epsilon)$. This function has an inverse satisfying the assumptions of Theorem 17.1.

DEFINITION 17.2: The equilibrium of (2.7) is said to be uniformly stable in the whole if the assumptions of Theorem 17.1 are satisfied for every arbitrarily large r_1.

Autonomous and periodic differential equations form a special case; this follows from:

THEOREM 17.2: *If the equilibrium of an autonomous, or periodic differential equation is stable, then it is also uniformly stable; stability in the whole implies uniform stability in the whole.*

Proof (Hahn): Because of the stability, the expression

$$\mu(0, \mathsf{x}_0, t_0) = \sup_{u \geq 0} |\mathsf{p}(t_0 + u, \mathsf{x}_0, t_0)|$$

(cf. (16.1)) is certainly bounded if the initial points x_0 are chosen from the interior of a fixed spherical domain $\mathfrak{K}_r$. Since the differential equation admits the period w (w is arbitrary in the autonomous case), the relation

$$\mathsf{p}(t + w, \mathsf{x}_0, t_0 + w) = \mathsf{p}(t, \mathsf{x}_0, t_0) \qquad (17.3)$$

holds. Consequently, μ is periodic with respect to t_0, and if r remains sufficiently small, the function

$$\rho(r) = \sup |\mathsf{p}(t_0 + u, \mathsf{x}_0, t_0)| \qquad (17.4)$$

can be formed. The variables have to be varied according to the conditions

$$u \geqq 0, \quad t_1 \leqq t_0 \leqq t_1 + w, \quad |\mathsf{x}_0| \leqq r \qquad (t_1 \geqq 0, \text{ arbitrary})$$

The function (17.4) satisfies the conditions of Theorem 17.1 for arbitrarily large r_1.

Massera [1, 4] was the first to formulate the theorem and to give an indirect proof.

In the case of asymptotic stability, uniformity refers to the limit process of Definition 2.4. Since the initial point and the initial instant both occur as parameters, two types of uniformity can be defined.

DEFINITION 17.3(a) (Antosiewicz [2]): *The equilibrium of the differential equation* (2.7) *is said to be* quasi-equiasymptotically *stable if there is for each initial instant t_0 a number $\delta = \delta(t_0)$ such that the solution $\mathsf{p}(t, \mathsf{x}_0, t_0)$ tends uniformly towards zero for $|\mathsf{x}_0| < \delta$ as $t \to \infty$.* If this is the case, there exists for each $\epsilon > 0$ a $\tau = \tau(\epsilon, t_0)$ such that

$$|\mathsf{p}(t, \mathsf{x}_0, t_0)| < \epsilon \quad \text{for} \quad t > t_0 + \tau, \, |\mathsf{x}_0| < \delta \qquad (17.5)$$

DEFINITION 17.3 (Massera [1]): *If the equilibrium is both stable and quasi-equiasymptotically stable, then it is said to be* equiasymptotically *stable.* In this case, one also uses the terminology *uniform asymptotic stability with respect to the spatial coordinates* (i.e., with respect to the phase space, or, in other words, with respect to the initial point) (Krasovskii [16]).

In the case of linear equations, equiasymptotic stability is a consequence of asymptotic stability. Massera [1, 4] gave examples showing that this conclusion is not always valid.

If for given ϵ the number τ can be chosen independently of t_0, then the equilibrium is uniformly equiasymptotically stable (Antosiewicz [2]). In addition to this, if the equilibrium is uniformly stable, one arrives at:

DEFINITION 17.4 (Malkin [20]): *The equilibrium of* (2.7) *is called uniformly asymptotically stable if*

1. *the equilibrium is uniformly stable,*
2. *for every $\epsilon > 0$ a number $\tau = \tau(\epsilon)$ depending only on ϵ, but not on the initial instant t_0, can be determined such that the inequality*

$$|\mathsf{p}(t, \mathsf{x}_0, t_0)| < \epsilon \qquad (t > t_0 + \tau)$$

holds, provided x_0 belongs to a spherical domain $\mathfrak{K}_\eta$ whose radius η is independent of ϵ.

To Theorem 17.1 there corresponds:

THEOREM 17.3: *Necessary and sufficient for the second assumption of Definition 17.4 is the existence of a function $\sigma(r)$ with the following properties:*

(a) $\sigma(r)$ *is defined, continuous, and monotonically decreasing, for all* $r \geq 0$,

(b) $\lim\limits_{r \to \infty} \sigma(r) = 0$,

(c) *provided the initial points belong to a fixed spherical domain* $\mathfrak{K}_\eta$, *the relation*

$$|\mathsf{p}(t, \mathsf{x}_0, t_0)| \leq \sigma(t - t_0) \tag{17.6}$$

holds.

The proof is similar to that of Theorem 17.1. With a given ϵ, the lower bound $\bar{\tau}(\epsilon)$ is formed for the numbers τ of Definition 17.4; a continuous monotonically decreasing function $\hat{\tau}(\epsilon) \geq \bar{\tau}(\epsilon)$ is chosen whose inverse function is taken as $\sigma(r)$. This function possibly depends on the initial point x_0, but can be uniformly estimated, so that (17.2) and (17.6) can be combined into a single inequality. Then one obtains:

THEOREM 17.4 (Hahn): *Necessary and sufficient for uniform asymptotic stability of the equilibrium is the existence of two functions $\kappa(r)$ and $\vartheta(r)$ with the following properties:*

(a) $\kappa(r)$ *satisfies assumptions (a) and (b) of Theorem 17.1,*

(b) $\vartheta(r)$ *satisfies the corresponding assumptions of Theorem 17.3;*

(c) *in addition, the inequality*

$$|\mathsf{p}(t, \mathsf{x}_0, t_0)| \leq \kappa(|\mathsf{x}_0|) \, \vartheta(t - t_0) \tag{17.7}$$

holds, provided the initial points x_0 *belong to a fixed spherical domain* $\mathfrak{K}_\eta$.

DEFINITION 17.5 (Barbašin and Krasovskii [2]): The equilibrium of differential equation (2.7) is said to be *uniformly asymptotically stable in the whole*, if the following two conditions are satisfied:

(a) The equilibrium is uniformly stable in the whole;

(b) for any two numbers $\delta_1 > 0$ and $\delta_2 > 0$ there exists a number $\tau(\delta_1, \delta_2)$ such that

$$|\mathsf{p}(t, \mathsf{x}_0, t_0)| < \delta_2$$

if $t \geq t_0 + \tau(\delta_1, \delta_2)$ and $|\mathsf{x}_0| < \delta_1$.

> *Remark:* In the case of asymptotic stability in the whole, the necessity of an inequality of the form (17.7) cannot be verified by the above argument. The sufficiency is evident.

As a counter-part to Theorem 17.2, we have:

THEOREM 17.5: *If the equilibrium of an autonomous or periodic differential equation is asymptotically stable, or asymptotically stable in the whole, then this asymptotic stability is uniform.*

Proof: According to the assumption, the expression $\mu(t, x_0, t_0)$, defined in (16.1), is periodic in t_0 and approaches zero for fixed initial values x_0, t_0 as $t \to \infty$. As in the proof of Theorem 17.2, it can be seen that the expression
$$\sigma(\tau) = \sup |\mathsf{p}(t_0 + u, x_0, t_0)|$$
represents for $u \geq \tau$, $t_1 \leq t_0 \leq t_1 + w$, $|x_0| \leq r$, a comparison function of the properties required by Theorem 17.3.

This theorem was also precisely formulated first by Massera [4], who provided an indirect proof. Gradštein [1] gave a proof for the autonomous case.

A sufficient condition for uniform stability is obtained by a thorough analysis of the proofs of Theorems 4.1 and 4.2. This analysis shows that the existence of a Liapunov function with the properties required by those theorems implies uniform stability. This leads to:

Theorem 17.6: *If there exists a positive definite decrescent Liapunov function v such that its total derivative $\dot{v}$ for (2.7) is negative semi-definite, then the equilibrium is uniformly stable. If $\dot{v}$ is negative definite, the equilibrium is uniformly asymptotically stable; if v is radially unbounded, the equilibrium is uniformly asymptotically stable in the whole.*

The theorem was established in several steps by K. P. Persidskii [5], Malkin [20], Barbašin and Krasovskii [2]. The following proof is due to Massera [4] (cf. also Kurzweil [1]).

Proof: We use the notations of (1.8), (1.9), and Sec. 4. The relation
$$\varphi(|x|) \leq v(x, t) \leq \psi(|x|) \qquad (17.8)$$
holds in $\Re_{h_1, t_0}$. Let $0 < \epsilon < h_1$ and $\zeta = \zeta(\epsilon)$ be chosen so that $\varphi(\epsilon) > \psi(\zeta)$. If $|x_0| < \zeta$, then
$$|\mathsf{p}(t, x_0, t_0)| < \epsilon \quad \text{for} \quad t > t_0$$
for, by assumption, $v(\mathsf{p}(t, x_0, t_0), t) \equiv v(t)$ does not increase with t. Thus
$$v(t) \leq v(t_0) \leq \psi(\zeta) < \varphi(\epsilon) \qquad (17.9)$$
According to (17.8), $v(x, t) \geq \varphi(\epsilon)$ for $|x| = \epsilon$. Hence, from (17.9) the inequality
$$|\mathsf{p}(t, x_0, t_0)| < \epsilon$$
can be inferred. This proves the uniformity of the stability.

Now, let $\dot{v}$ be negative definite. For $\epsilon = h_1$, let $\bar{\delta}$ denote the number $\delta(\epsilon)$ of Definition 17.1, and let $\overline{w}_0$ be the minimum of $|\mathsf{p}(t, x_0, t_0)|$ in the interval $[t_0, t]$. Then the inequality
$$v(t) \leq v(t_0) - (t - t_0)\chi(\overline{w}_0) < \psi(\bar{\delta}) - (t - t_0)\chi(\overline{w}_0) \qquad (17.10)$$
follows from the integral estimate (4.5). Let

$$\tau = \frac{\psi(\bar{\delta})}{\chi(\zeta(\epsilon))}$$

where $\zeta(\epsilon)$ was defined above. If the relation $\overline{w}_0 \geqq \zeta(\epsilon)$ were true, then from (17.10) for $t = t_0 + \tau$ the inequality

$$v(t_0 + \tau) \leqq \psi(\bar{\delta}) - \tau\chi(\overline{w}_0) \leqq 0$$

would follow which contradicts the positivity of v. Hence, in the interval $(t_0, t_0 + \tau)$ there is a point t_1 for which

$$|\mathsf{p}(t_1, \mathsf{x}_0, t_0)| < \zeta(\epsilon)$$

Then, certainly, $|\mathsf{p}(t, \mathsf{x}_0, t_0)| < \epsilon$ for $t > t_1$ and, even more so, for $t > t_0 + \tau$. However, the number τ does not depend on t_0; i.e., the asymptotic stability is uniform. If the equilibrium is stable in the whole, or asymptotically stable in the whole, respectively, the consideration is valid for each h and furnishes the last statement of the theorem.

Remarks:

1. The assumptions of Theorem 17.6 about v can be modified. For instance, if v is positive definite, and decrescent, if (4.6) holds, and if the equilibrium of the differential equation introduced in Sec. 4, Remark 5, is uniformly stable, or uniformly asymptotically stable, then the corresponding statement is valid for the equilibrium of (2.7) (Corduneanu [3]). If a bounded function satisfies the conditions established for the special case of Remark 5, Sec. 4, then uniform asymptotic stability of the equilibrium is a consequence of uniform stability (Massera [4]).

2. The assumption of Sec. 4, Remark 4, assures uniform stability of the equilibrium with respect to the phase space coordinates (Krasovskii [15, 16]). Massera [1], and S. K. Persidskii [3, 4] established some other sufficient conditions. In these conditions, Liapunov functions are used for which the properties "positive definite," and "decrescent" are somewhat weakened.

3. The fact that the assumptions of Theorem 4.2 already assure uniform asymptotic stability yields an explanation for the many possibilities of weakening these assumptions (cf. the Remarks of Sec. 4). However, the conditions of Theorem 4.2 are necessary and sufficient for uniform asymptotic stability (cf. Sec. 18).

4. The equivalence of stability and uniform stability, expressed by Theorems 17.1 and 17.2 in the autonomous, or periodic case, does not hold for any arbitrary differential equation. For instance, the differential equation

$$\dot{x} = -x/(1 + t) \tag{17.11}$$

has the general solution

$$p(t, x_0, t_0) = x_0(1 + t_0)/(1 + t)$$

which does not tend towards zero uniformly with respect to t_0. Nor does equiasymptotic stability generally follow from uniform stability, and uniform asymptotic stability does not necessarily follow from uniform stability

(Massera [1, 4]). Thus, the introduced refinements of the concept of stability are truly refinements.

Differential equations with almost-periodic coefficients are sometimes encountered in investigations of oscillation problems. For example, such is the case if the differential equation of the perturbed motion is formed for the almost periodic solution of an autonomous differential equation [cf. (2.6)]. In this connection, the following problem arises which is still unsolved.

Problem: Is it true to state that for differential equations with almost-periodic coefficients (asymptotic) stability of the equilibrium is equivalent to uniform (asymptotic) stability?

5. A modification introduced by Yoshizava [1] ("equistable") is only of interest for differential equations whose solutions are not uniquely determined by the initial values. For such differential equations, Kurzweil [3] defined a concept which comprises uniform asymptotic stability in the whole. Let $\mathfrak{G}$ be an open subset of the phase space containing the origin, and let $\mathfrak{F}$ be its complement. Let $\rho(x, \mathfrak{F})$ be the distance of x from $\mathfrak{F}$; furthermore, let

$$\omega(x) = \max\left(|x|, \frac{1}{\rho(x, \mathfrak{F})} - \frac{2}{\rho(0, \mathfrak{F})}\right)$$

The equilibrium is called *strongly stable* in $\mathfrak{G}$ if for arbitrarily given positive numbers β and ϵ there exist two positive numbers $b = b(\beta)$ and $\tau = \tau(\beta, \epsilon)$ with the following properties:

(a) $b(\beta)$ tends monotonically towards zero with $\beta \to 0$;

(b) if $q(t)$ is a solution of the differential equation, defined in the interval $t_0 \leq t \leq t_1 < \infty$, and if $q(t)$ satisfies the inequality $\omega(q(t_0)) \leq \beta$, then there exists a solution $p(t)$ defined for all $t \geq t_0$ which coincides with $q(t)$ in the interval (t_0, t_1) and which satisfies the conditions

$$\omega(p(t)) < b \quad \text{for} \quad t > t_0$$
$$\omega(p(t)) < \epsilon \quad \text{for} \quad t \geq t_0 + \tau$$

6. As a consequence of the definition of uniform asymptotic stability, a lower bound for the rate, so to speak, at which motions approach the equilibrium is introduced. By estimating this rate above, another refinement of the stability concept is obtained. According to Zubov [6, 7], the equilibrium of differential equation (2.7) is said to be *uniformly attracting* if it is asymptotically stable and if for every number ζ satisfying the inequality $0 < \zeta < \eta$ (η is the number introduced in Definition 17.3) numbers $\tau > 0$ and $\alpha > 0$ can be found such that for all $t_0 > 0$ and all t of the interval $t_0 \leq t \leq t_0 + \tau$ the inequality

$$|p(t, x_0, t_0)| > \alpha$$

holds, provided

$$\zeta \leq |x_0| \leq \eta$$

The definition can be expressed in analogy to Theorem 17.5 by means of the two comparison functions $\kappa_1(r)$ and $\vartheta_1(r)$ which must satisfy assumptions (a) and (b) of Theorems 17.11 and 17.3, respectively: The equilibrium which is assumed to be asymptotically stable is uniformly attracting if and only if

$$|\mathsf{p}(t, \mathsf{x}_0, t_0)| > \kappa_1(|\mathsf{x}_0|)\,\vartheta_1(t - t_0) \qquad (17.12)$$

provided the initial points belong to the interval $\zeta \leq |\mathsf{x}_0| \leq \eta$.

The inequalities (17.7) and (17.12) hold simultaneously if the equilibrium is both uniformly stable and uniformly attracting. These inequalities then guarantee that the motions remain in a "tube-like" domain.

7. A refinement of the concept of uniform asymptotic stability, the so-called *exponential stability*, will be dealt with in Sec. 22 along with a discussion of the corresponding refinements of the concept of instability.

18. THE INVERSION OF THE STABILITY THEOREMS

(a) Let the differential equation

$$\dot{\mathsf{x}} = \mathsf{f}(\mathsf{x}, t) \qquad (18.1)$$

be given, and suppose the stability behavior of the equilibrium in the sense of the definitions of Sec. 2, or Sec. 17, to be known. The question is whether or not there exists a Liapunov function v which allows the application of the theorems of Secs. 4 and 17. The various existence theorems, which are also called converse theorems of the direct method, differ in several respects, namely:

(α) in the assumptions about the right hand side of differential equation (18.1);

(β) in the assumptions about the stability of the equilibrium;

(γ) in the statements on the Liapunov function v, particularly with respect to existence and continuity of its partial derivatives.

(b) *Converse Theorems on Weak Stability.*

THEOREM 18.1 (Kurzweil [1]): Let the functions $f_i(\mathsf{x}, t)$ and $\partial f_i/\partial x_j$ be continuous in a domain $\mathfrak{K}_{h,t_0}$ and let the equilibrium of (18.1) be stable. Then there exists a positive definite function $v(\mathsf{x}, t)$ which has in $\mathfrak{K}_{h,t_0}$ continuous first partial derivatives with respect to all variables, and whose total derivative $\dot{v}$ for (18.1) is negative semi-definite.

Proof: As a consequence of the assumption of stability, a number $\delta > 0$ can be chosen such that the inequality $|\mathsf{p}(t, \mathsf{x}_0, t_0)| < h/2$ follows from $|\mathsf{x}_0| < \delta$ for $t \geq t_0$. Let $\mathfrak{G}_1$ and $\mathfrak{G}_2$ be the sets of those points (x, τ) of $\mathfrak{K}_{h,t_0}$ for which $|\mathsf{p}(t_0, \mathsf{x}, \tau)| < \delta/2$, and $|\mathsf{p}(t_0, \mathsf{x}, \tau)| < \delta$, respectively, and let $\tilde{\mathfrak{G}}_1$ be the complement of $\mathfrak{G}_1$ in $\mathfrak{K}_{h,t_0}$. Let $\gamma(r)$ denote a function which vanishes for $r \leq 0$, which is positive and continuously differentiable for $r > 0$, and which takes on the constant value $\gamma(r) = \delta^2$ for $r \geq \delta/2$. The Liapunov function is now defined in the following manner:

$$v(\mathsf{x}, \tau) = \delta^2 \qquad ((\mathsf{x}, \tau) \in \tilde{\mathfrak{G}}_1)$$
$$v(\mathsf{x}, \tau) = \gamma(|\mathsf{p}(t_0, \mathsf{x}, \tau)|) \qquad ((\mathsf{x}, \tau) \in \mathfrak{G}_2)$$

The two definitions agree in $\mathfrak{G}_1 \cap \mathfrak{G}_2$. The function v is nonnegative, and it vanishes at the origin. It is continuously differentiable in $\mathfrak{K}_{h,t_0}$, and its total derivative for (18.1) is identically equal to zero, since v is constant along every motion. The fact that v is positive definite can be shown as follows: Let $0 < \epsilon < h$ and $\bar{\delta} < \delta$ be so determined that the inequality $|\mathsf{p}(t, \mathsf{x}, t_0)| < \epsilon$ follows from $|\mathsf{x}| < \bar{\delta}$ for $t > t_0$. If $\epsilon \leq |\mathsf{x}| \leq h$ and $(\mathsf{x}, \tau) \in \mathfrak{G}_2$, then certainly $|\mathsf{x}_1| \geq \bar{\delta}$ for $\mathsf{x}_1 = \mathsf{p}(t_0, \mathsf{x}, \tau)$, since $\mathsf{p}(t, \mathsf{x}_1, t_0) = \mathsf{p}(t, \mathsf{x}, \tau)$. It follows that

$$v(\mathsf{x}, \tau) \geq \min(\delta^2, \gamma(|\bar{\delta}|))$$

The essential parts of Theorem 18.1 have been proved already by K. P. Persidskii [5]. His result differs only in that the domain for which the existence of v is assured forms in general only a subdomain of $\mathfrak{K}_{h,t_0}$. Persidskii showed the following: If the solution $\mathsf{p}(t, \mathsf{x}_0, 0)$ is solved with respect to x_0 and if x_0 is put equal to $\mathsf{q}(t, \mathsf{p})$, then the expression $|\mathsf{x}_0|^2 = v(\mathsf{p}, t)$ can be used as a Liapunov function for (18.1). A Liapunov function which is continuous at the origin, but not necessarily differentiable, was constructed by Yoshizava [1] in an even simpler manner under the sole assumption of continuity of f.

The following Theorem is somewhat sharper than Theorem 18.1.

THEOREM 18.2 (Kurzweil [1], Krasovskii [12]): Suppose that in addition to the assumptions of Theorem 18.1 the stability is uniform. Then a positive definite, decrescent function $v(\mathsf{x}, t)$ exists in each domain $\mathfrak{K}_{h_1,t_0}$ with $0 < h_1 < h$ which has continuous first derivatives in $\mathfrak{K}_{h_1,t_0}$ with respect to all variables, and the total derivative of v for (18.1) is negative semidefinite.

The proof of this theorem according to Kurzweil [1] is more extensive than that of Theorem 18.1. It is best given in several steps. The first step is to show that the auxiliary function $m(\mathsf{x}, t) = \min |\mathsf{p}(\tau, \mathsf{x}, 0)|$ $(0 \leq \tau \leq t)$ is positive definite and decrescent by using primarily the uniformity of the stability. This function is then "smoothed," i.e., it is replaced by a differentiable function $\tilde{m}(\mathsf{x}, t)$ whose values are only slightly different from those of m. In the next step, a function is formed which has the additional property of being decreasing along each motion. This function, constructed by means of suitable estimates for the partial derivatives of $\tilde{m}$, is the desired Liapunov function. In connection with Theorem 17.6 it then follows:

The conditions of Theorem 18.1 are necessary and sufficient for uniform stability.

Krasovskii [12] proved Theorem 18.2 by assuming the boundedness of the partial derivatives $\partial f_i/\partial x_j$ and by using a topological lemma of Barbašin [1]. Kurzweil and Vrkoč [1] showed that the assumption of continuity of the partial derivatives may be replaced by other weaker conditions. They have furthermore demonstrated that, in the case of only continuous right hand sides, uniformity of stability alone is not sufficient to assure the existence of a continuous Liapunov function, but that additional assumptions about f(x, t) are needed. However, if a continuous positive definite function v with nonnegative derivative $\dot v$ exists at all, then there also exists a function v^* with the same properties and which has partial derivatives of any order. If v is decrescent, the same holds for v^*. It should be mentioned that a time-independent Liapunov function which satisfies the conditions of Theorem 4.1 cannot be constructed for every autonomous differential equation with a stable equilibrium. The situation is different in the case of uniform asymptotic stability.

(c) *Converse Theorems on Asymptotic Stability.*

THEOREM 18.3 (Massera [4]): *Let the equilibrium of* (18.1) *be uniformly asymptotically stable, and let* $f \in C_0$ *in* $\mathfrak{R}_{h,t_0}$. *Then there exists a positive definite decrescent function v whose total derivative for* (18.1) *is negative definite, and which has partial derivatives of any order.*

Corollary 1: If $f \in \bar{C}_0$ (i.e., if there exists a uniform Lipschitz constant), then v can be determined such that all partial derivatives are uniformly bounded in $\mathfrak{R}_{h,t_0}$ with respect to t and such that the same bound is applicable to all derivatives.

Corollary 2: If the differential equation is periodic, or autonomous, v can be constructed to be independent of t, or to be periodic in t, respectively.

The proof of this important theorem, which actually answers the principal question of the inversion problem, is an elaboration of a proof developed by Massera [1] for a special case (assumptions: f autonomous, or periodic, $f \in C_1$; statement: $v \in C_1$). Massera used the following:

Lemma: Let $\rho(r)$ and $\sigma(r)$ be continuous and positive functions defined for $r \geqq 0$, the first of which increases with r, the second of which tends towards zero as $r \to \infty$. Then there exists a function $\omega(r) \in K$ defined for all $r \geqq 0$ (cf. Sec. 1) with the following properties: (a) the derivative $\omega'(r)$ exists and belongs also to the class K; (b) the integrals

$$\int_0^\infty \omega(\sigma(r))\, dr, \qquad \int_0^\infty \omega'(\sigma(r))\, \rho(r)\, dr$$

are bounded.

Malkin [20] proved a somewhat weaker form of Theorem 18.3 by means of Massera's method (assumption: $f \in C_1$; statement: $v \in C_1$). In particular, from uniform stability he first inferred the existence of two functions $\rho(r)$ and $\sigma(r)$ satisfying the lemma and permitting the estimates

$$\left|\frac{\partial |p(t_0 + r, x, t_0)|^2}{\partial x_i}\right| < \rho(r), \qquad \left|\frac{\partial |p(t, x, t_0)|^2}{\partial t_0}\right|_{t=t_0+r} < \rho(r)$$

$$|p(t_0 + r, x, t_0)| < \sigma(r)$$

[$\sigma(r)$ is the comparison function of Theorem 17.3]. If $\omega(r)$ is the function of the lemma corresponding to $\rho(r)$ and $\sigma(r)$, then the expression

$$v(\mathsf{x}, t) = \int_t^\infty \omega(|\mathsf{p}(\tau, \mathsf{x}, t)|)\, d\tau$$

satisfies the assumptions of Theorem 4.2.

Because of the weak assumptions of Theorem 18.3, Massera [4] had to proceed somewhat differently. He began with the function $p(\mathsf{x}, t) = \sup_{\tau \geq 0} |\mathsf{p}(t + \tau, \mathsf{x}, t)|$ defined in the motion space (with different notation, the function was used already in the proof of Theorem 17.2). He then constructed an auxiliary function $g(r)$ of the real variable r, which satisfies certain conditions imposed on its rate of change. This function is primarily determined by the rate with which $p(\mathsf{x}, t)$ varies in dependence upon (x, t). A definite, decrescent Liapunov function $u(\mathsf{x}, t)$ which has a negative definite derivative can then be defined by means of the function $g(p(\mathsf{x}, t))$, which again has to be considered as a function in the motion space. An ingenious smoothing process finally transforms $u(\mathsf{x}, t)$ into a function $v(\mathsf{x}, t)$ which satisfies all statements.

Barbašin [1] approached the problem by means of completely different tools; he used methods of the theory of dynamic systems (cf. Sec. 20). He first showed the existence of a Liapunov function v with $v \in C_m$ in the case of autonomous differential equations provided that $\mathsf{f} \in C_m$. However, the result can be immediately extended to certain nonautonomous differential equations. Another, somewhat shorter, proof of Barbašin's result was given by Krasovskii [19].

The additional assumption of uniformity of the asymptotic stability in the whole becomes necessary for the converse statement of Theorem 4.3.

THEOREM 18.4 (Massera [4]): If the equilibrium of (18.1) is uniformly asymptotically stable in the whole and if $\mathsf{f} \in C_0$, then there exists a positive definite, decrescent, and radially unbounded function v with negative definite derivative. The function v has partial derivatives of any order.

Most parts of the proof are identical with the proof of Theorem 18.3. Earlier, Barbašin and Krasovskii [1] formulated Theorem 18.4 for autonomous motions with reference to Barbašin's [1] lemma mentioned above in connection with the proof of Theorem 18.2. However, as Erugin [6] noted, the assumption must be made that the solutions can be continued into the interval $-\infty < t < t_0$, and this is not always admissible. (An additional critical remark of Erugin [6, 8] has been considered by Barbašin and Krasovskii [2].) Theorem 18.4 was proved with $\mathsf{f} \in C_1$ and $v \in C_1$ in a paper of Barbašin and Krasovskii [2].

Kurzweil [2, 3] restricted the assumptions of Theorem 18.3 to the extent that $\mathsf{f}(\mathsf{x}, t)$ needs only be continuous. He replaced the requirement of uniform asymptotic stability by strong stability as defined in Sec. 17, Remark 5. Kurzweil's proof was established independently of Massera's proof, but it is based on the same procedure. However, it is considerably longer, since the weaker assumptions and the nonuniqueness of the solutions necessitate a number of additional considerations.

Krasovskii [20] proved an existence theorem for a decrescent Liapunov function with definite derivative. It was found that "noncritical and uniform" behavior of the trajectories are necessary as well as sufficient conditions. This means, for any two sufficiently small numbers $\delta > 0$ and $\eta > 0$ there exists a number $\tau = \tau(\delta, \eta)$ such that every motion $p(t, x_0, t_0)$ with $|x_0| > \eta$ reaches the hyper-surface $|x| = \delta$ along at least one half-trajectory, i.e., for $t \geq t_0$, or $t \leq t_0$, respectively, before the end of the time period $|t - t_0| < \tau$. Therefore, the motions are not allowed to approach the t-axis of the motion space arbitrarily closely, or to leave the t-axis arbitrarily slowly. In the stable case, the so defined concept coincides with the concept of uniform asymptotic stability. The theorems on equiasymptotic stability, mentioned in Sec. 17, Remark 2, can also be converted (Massera [5], S. K. Persidskii [3, 4], Antosiewicz [2]).

(d) Converse theorems with the assumption of asymptotic stability of the equilibrium without uniformity are still unknown. However, Zubov [6] proved some theorems in which a more general statement on the rate of decay of the motion replaces the inequality (17.7). Two auxiliary functions $\varphi(t)$ and $\psi(t)$ which are positive and continuous for $t \geq 0$ are needed to formulate these theorems. The integral

$$J(t, t_0) = \int_{t_0}^{t} \frac{\psi(\tau)}{\varphi(\tau)} d\tau$$

shall be bounded for $t \geq t_0 \geq 0$ and it shall approach infinity with increasing t. Furthermore, the expression

$$\kappa(t, t_0; \alpha, \beta) = \left(\frac{\varphi(t_0)}{\varphi(t)}\right)^{1/(1+\alpha)} \exp(-\beta J(t, t_0))$$

depending on two positive parameters α and β, shall approach zero as $t \to \infty$. With these notations, the following theorem holds:

THEOREM 18.5 (Zubov [6]): Let the equilibrium of (18.1) be asymptotically stable and let

$$l_1 \kappa(t, t_0; \alpha, \beta_1) |x_0|^2 \leq |p(t, x_0, t_0)|^2 \leq l_2 \kappa(t, t_0; \alpha, \beta_2) |x_0|^2$$

for all sufficiently small values of $|x_0|$; l_1 and l_2 are certain positive constants. The existence of two functions

$$v = |x|^{2\alpha} \varphi(t) x^T A(x, t) x$$

and

$$w = |x|^{2\alpha} \psi(t) x^T B(x, t) x$$

with the properties given below is necessary and sufficient for such behavior of the solutions:
  (a) they are defined and continuous in a domain $\mathfrak{K}_{h,t_0}$;
  (b) for sufficiently small $|x|$, the relations

$$a_1|x|^2 \leq x^T A(x, t) x \leq a_2|x|^2 \qquad (a_1 > 0, a_2 > 0)$$
$$b_1|x|^2 \leq x^T B(x, t) x \leq b_2|x|^2 \qquad (b_1 > 0, b_2 > 0)$$

hold;
  (c) $\dot{v} = -w$

Other expressions with similar properties may be used instead of the expression denoted by κ.

19. THE INVERSION OF THE INSTABILITY THEOREMS

The converse statements of the instability theorems are generally less intricate than those on asymptotic stability, mainly because the property of being unstable is considerably less characteristic than that of being asymptotically stable. The fact that the converse theorems on stability were proved earlier than the corresponding theorems on instability does not mean that the latter theorems are more complicated. The requirements contained in Theorem 5.2 are more stringent than those of Theorem 5.1. The assumptions of Theorem 5.2 comprise a statement on the behavior of the derivative in a whole neighborhood of the origin; the assumptions of Theorem 5.1 (and also those of Theorem 5.3) require knowledge of v only in a domain for which the origin is a boundary point. Correspondingly, the converse of Theorem 5.2 provides a more general statement, but also requires stronger assumptions.

The first important result was obtained by Krasovskii [8]; it refers to the autonomous differential equation

$$\dot{x} = f(x) \qquad (19.1)$$

and is based on a lemma (which is in itself of interest) on the existence of a Liapunov function v with *definite* derivative $\dot{v}$ (cf. also Krasovskii [14]).

THEOREM 19.1 (Krasovskii [8]): The existence of a neighborhood of the origin of the phase space, not containing a complete phase trajectory, is necessary and sufficient for the existence of a Liapunov function v whose derivative $\dot{v}$ for (19.1) is definite.

Using the results of Barbašin [1] (cf. proof of Theorem 18.2), Krasovskii proved this theorem by applying topological methods, and by means of a construction procedure similar to that of Massera [1]. The assumption of Theorem 19.1 for example, is not satisfied in the critical cases of Sec. 8; it is also not satisfied for the motion described by (2.11), since every neighborhood of the origin contains closed phase curves (cf. also Krasovskii [20] and Sec. 18 (c)). Krasovskii [24] formulated the property of the motion which is characteristic for the existence of a Liapunov function with definite derivative also in the nonautonomous case. A sequence of spherical domains $\Re_k$ is considered whose radii h_k approach zero monotonically. Let $\mathfrak{H}$ be any subdomain of $\Re_{h,t_0}$ (h fixed). It has to be required that a number T_k can be chosen with the following property: That part of the motion $\mathsf{p}(t, \mathsf{x}_0, t_0)$ which is determined by $t_0 - T_k \leqq t \leqq t_0 + T_k$ is positively not completely contained in $\mathfrak{H}$, provided the initial point x_0 lies in the exterior of the domain $\Re_k$, and provided, of course, that t_0 is so large that the motion can be continued sufficiently far into the domain $t < t_0$ (cf. also Sec. 4, Remark 3).

Krasovskii [8] proved the following theorem by means of Theorem 19.1.

THEOREM 19.2: Let the equilibrium of (19.1) be unstable. The existence of a neighborhood of the origin not containing a complete phase trajectory of (19.1) is necessary and sufficient for the existence of a function v which can take on negative values in every neighborhood of the origin and whose total derivative $\dot{v}$ for (19.1) is negative definite.

Theorem 19.2 contains Theorem 5.2 in the case of autonomous motions and the conditions for its inversion. Using the above mentioned extension of Theorem 19.1, Krasovskii [24] proved the converse of Theorem 5.2 in the nonautonomous case.

The converse of Chetaev's Theorem 5.1 can be accomplished without additional assumptions about the integral curves of the original equation.

THEOREM 19.3 (Vrkoč [1]): Let the equilibrium of the differential equation

$$\dot{\mathsf{x}} = \mathsf{f}(\mathsf{x}, t) \qquad (\mathsf{f} \in C_1) \qquad (19.2)$$

be unstable. Then there exists a Liapunov function v in a certain domain $\Re_{h,t_0}$ with the following properties:

(a) $v \in C_1$;

(b) there exists a domain $v < 0$ (cf. Theorem 5.1) where v is bounded; and

(c) $\dot{v} = v$

The relatively simple proof consists of a transfinite construction procedure for the Liapunov function on which even the additional condition may be imposed to be negative in a certain *domain of instability* and only there. This domain of instability $\mathfrak{G}(\epsilon)$ is defined for given $\epsilon > 0$ and fixed t_0 as follows: Let the initial points $\tilde{\mathsf{x}}_0$ be such that

$$\max |\tilde{x}_{i0}| < \epsilon, \quad \max |p_i(t, \tilde{\mathsf{x}}_0, t_0)| > \epsilon \qquad (t > t_0)$$

$\mathfrak{G}(\epsilon)$ is the set of all these points $\tilde{\mathsf{x}}_0$.

Theorem 19.3 provides an inversion of Theorem 5.3, from which the equivalence of Theorems 5.1 and 5.3 (mentioned already in Sec. 5) follows. Another inversion of Theorem 5.3 was found by Krasovskii [19].

Theorem 5.4 can also be converted.

THEOREM 19.4 (S. K. Persidskii [1]): If the equilibrium is completely unstable, there exists in a domain $\Re_{h,t_0}$ a positive function v with nonnegative derivative; v tends uniformly with respect to x towards zero with increasing t.

The proof is accomplished in a manner similar to K. P. Persidskii's [5] proof of Theorem 18.1. In the theorems of Liapunov and Chetaev (Sec. 5) the property of being "not stable" is characterized by means of Liapunov functions. Blichevskii [1] noted that the properties of being "not uniformly stable" or "not asymptotically stable," may as well be characterized by suitable Liapunov functions.

20. ON THE STABILITY THEORY OF DYNAMICAL SYSTEMS

Some of the theorems given in the previous section assume a particularly elegant form if they are formulated as statements on dynamical systems. In fact, one can even develop a large part of the theory of the direct method immediately by such an approach. This approach was taken by Zubov [6] whose presentation and terminology will be used in this section. We state only results.

Let a vector function q(x, t) be given in the n-dimensional phase space with the following properties:

(*a*) For each finite x, the components of q are defined in the entire interval $-\infty < t < +\infty$ as one-valued functions of their arguments;
(*b*) q(x, 0) = x;
(*c*) q(q(x, t_1), t_2) = q(x, $t_1 + t_2$).

Then the function q(x, t) is said to define a *dynamical system* in the phase space. The set of the points q(x, t) for fixed x is called a *trajectory* of the system. The system can then be considered either as the set of its trajectories which determine a schlicht coverage of the phase space, or as a one-parameter group of continuous mappings of the phase space onto itself. In the latter case, the notation

$$x \to q(x, t)$$

is used.

A point set $\mathfrak{M}$ of the phase space, consisting of trajectories only, is called an *invariant set* of the system. If $x \in \mathfrak{M}$, it follows that q(x, t) $\in \mathfrak{M}$ for all t.

As usual, let

$$\rho(x, \mathfrak{M}) = \inf_{y \in \mathfrak{M}} |x - y| \qquad (20.1)$$

be the *distance* of the point x from the set $\mathfrak{M}$. The set of all points x with $0 < \rho(x, \mathfrak{M}) < r$ represents a *neighborhood* $\mathfrak{U}(\mathfrak{M}, r)$ of the set $\mathfrak{M}$. This definition, used by Zubov, differs somewhat from the usual definition.

A stability theory can now be based on the following definitions:

DEFINITION 20.1: A closed invariant set $\mathfrak{M}$ of a dynamical system is called *stable*, if for every $\epsilon > 0$ a number $\delta > 0$ can be found such that for all $t > 0$

$$\rho(q(x, t), \mathfrak{M}) < \epsilon$$

provided

$$\rho(x, \mathfrak{M}) < \delta$$

If, in addition,

$$\lim_{t \to \infty} \rho(q(x, t), \mathfrak{M}) = 0$$

then $\mathfrak{M}$ is said to be *asymptotically stable*.

DEFINITION 20.2: The closed invariant set $\mathfrak{M}$ is said to be *uniformly asymptotically stable*, if there exists a number δ_0 with the following property: For each $\epsilon > 0$ there exists a number $\tau = \tau(\epsilon) > 0$ such that

$$\rho(q(x, t), \mathfrak{M}) < \epsilon \quad \text{for} \quad t > \tau$$

provided

$$\rho(x, \mathfrak{M}) < \delta_0$$

DEFINITION 20.3: The invariant set $\mathfrak{M}$ is called *unstable* if there exists an $\epsilon > 0$ with the following property: For every $\delta > 0$, there is in $\mathfrak{U}(\mathfrak{M}, \delta)$ at least one point x_0 and there is an instant $t_1 > 0$ such that

$$\rho(q(x_0, t_1), \mathfrak{M}) \geq \epsilon$$

The refinement achieved by the transition from asymptotic stability to uniform asymptotic stability can be interpreted topologically: An asymptotically stable, closed, invariant set which has a compact neighborhood, is a priori uniformly asymptotically stable. For those properties defined in Definitions 20.1 and 20.2, conditions can be established which are completely analogous to those of the direct method; they are even both necessary and sufficient. In the following four theorems, $\mathfrak{M}$ denotes an invariant, closed, set of the given dynamical system $q(x, t)$, and $\mathfrak{U} = \mathfrak{U}(\mathfrak{M}, r)$ denotes a sufficiently small neighborhood of $\mathfrak{M}$.

THEOREM 20.1: Necessary and sufficient for stability of $\mathfrak{M}$ is the existence of a function $v(x)$ with the properties:

(a) $v(x)$ is defined in $\mathfrak{U}$;

(b) to every sufficiently small number $c_1 > 0$ there corresponds a number $c_2 > 0$ such that $v(x) > c_2$ if $x \in \mathfrak{U}$ and $\rho(x, \mathfrak{M}) > c_1$;

(c) to every sufficiently small number $d_2 > 0$ there corresponds a number $d_1 > 0$ such that $v(x) < d_2$ if $x \in \mathfrak{U}$ and $\rho(x, \mathfrak{M}) < d_1$;

(d) the function $v(q(x, t))$ does not increase for $t \geq 0$, as long as $q(x, t)$ remains in $\mathfrak{U}$.

THEOREM 20.2: Necessary and sufficient for asymptotic stability of $\mathfrak{M}$ is the existence of a function $v(x)$ which satisfies the assumptions of Theorem 20.1 and, in addition,

(e) $$\lim_{t \to \infty} v(q(x, t)) = 0 \qquad (20.2)$$

for $q(x, t) \in \mathfrak{U}$ and $t > 0$.

THEOREM 20.3: Necessary and sufficient for the uniformity of the asymptotic stability assured by Theorem 20.2 is the additional assumption:

(f) a neighborhood $\mathfrak{U}(\mathfrak{M}, \delta_1)$ exists such that the limit process (20.2) holds uniformly with respect to $x \in \mathfrak{U}(\mathfrak{M}, \delta_1)$.

THEOREM 20.4: Necessary and sufficient for the instability of $\mathfrak{M}$ is the existence of a function $v(x)$ with the properties:

(a) v is defined and bounded in $\mathfrak{U}$;
(b) in every arbitrarily small neighborhood of $\mathfrak{M}$, there exist points x for which $v(x) > 0$;
(c) for every point x, the equation $\dot{v} = gv + w$ is valid, where

$$\dot{v}(x) = \lim_{t \to 0+} \frac{v(q(x, t)) - v(x)}{t}$$

g is a positive number and w is a nonnegative function defined in $\mathfrak{U}$.

The transition from the stability theory of dynamical systems (of which only the most important theorems have been mentioned) to the stability theory of differential equations is accomplished by means of the following theorems.

THEOREM 20.5: To every autonomous differential equation

$$\dot{x} = f(x) \qquad (f \in E) \qquad (20.3)$$

there can be associated a dynamical system defined in an n-dimensional space such that the solutions of the differential equation correspond to the trajectories of the system and are equivalent to them with respect to the stability behavior.

THEOREM 20.6: If the equilibrium of a nonautonomous differential equation

$$\dot{x} = f(x, t) \qquad (f \in C_1 \text{ in } \mathfrak{K}_{h,0}, f(0, t) \equiv 0) \qquad (20.4)$$

is uniformly stable, then there can be associated to the differential equation a dynamical system, defined in an $(n + 1)$-dimensional space, for which the t-axis is a stable invariant set. If the equilibrium of (20.4) is uniformly asymptotically stable, then the t-axis is an asymptotically stable invariant set of the dynamical system.

If the right hand side of (20.3) belongs to the class E in the whole phase space and if the solutions $\mathsf{x}(t)$ are defined for all t ($-\infty < t < +\infty$), then the trajectories of the differential equation already form a dynamical system in the phase space. Otherwise, the function $\mathsf{f}(\mathsf{x})$ must be continued in some suitable manner into the whole phase space. For example, a new independent variable s can be introduced by the expression

$$ds/dt = \sqrt{1 + |\mathsf{f}|^2} \qquad (20.5)$$

The solutions are then defined for all values of s.

A corresponding procedure is applied in the nonautonomous case. The dynamical system is defined in the motion space and the variable s, introduced by (20.5), plays the role of the system parameter.

The principal statements of Theorems 4.1 to 4.3, 5.1, 18.1 to 18.3, and 19.3, can be obtained from Theorems 20.1 and 20.4 in conjunction with Theorems 20.5 and 20.6. The assumptions (b) and (c) of Theorem 20.1 generalize the definiteness, or the decrescentness, of the Liapunov function, respectively. Theorem 20.6 clarifies the meaning of uniformity of stability with respect to the inversion problem.

As already mentioned (Sec. 18), the proof of the converse theorem contained in Theorem 20.3 was first accomplished by Barbašin [1].

The theory of dynamical systems is not sufficient for the treatment of nonautonomous differential equations with nonuniformly stable equilibrium. Therefore, "general systems" are considered for which a stability theory can be established which is analogous to a certain degree to the theory sketched above (cf. Sec. 35, Zubov [6] and Lefschetz [2]).

21. ZUBOV'S METHOD OF CONSTRUCTION

Zubov ([1 to 6]) reduced the problem of constructing a Liapunov function for a given differential equation to the solution of a partial differential equation. This problem then becomes, in many cases, amenable to a systematic treatment. Furthermore, the method permits precise determination of the boundary of the domain of attraction. Zubov [3] first considered the stability of the equilibrium of the autonomous differential equation

$$\dot{\mathsf{x}} = \mathsf{f}(\mathsf{x}) \qquad (\mathsf{f}(0) = 0, \mathsf{f} \in E) \qquad (21.1)$$

He then extended the construction method to invariant sets of dynamical systems (cf. Sec. 20) with the result that it can also be applied to some classes of nonautonomous differential equations. In the meantime, the investigations have been summarized and presented in a monograph (Zubov [6]). The author developed a systematic theory of the direct method and succeeded in deriving a series of results by means of his construction approach. The core of Zubov's theory is:

THEOREM 21.1 (Zubov [3, 6]): Let $\mathfrak{A}$ be an open domain of the phase space and let $\bar{\mathfrak{A}}$ be its closure; $\mathfrak{A}$ shall contain the origin. Necessary and sufficient for $\mathfrak{A}$ to be the exact domain of attraction of the equilibrium of (21.1) is the existence of two functions $v(x)$ and $\varphi(x)$ with the following properties:

(a) $v(x)$ is defined and continuous in $\mathfrak{A}$, $\varphi(x)$ is defined and continuous in the whole phase space;
(b) $\varphi(x)$ is positive definite for all x;
(c) $v(x)$ is negative definite, and for $x \in \mathfrak{A}$, $x \neq 0$, the inequality $-1 < v(x) < 0$ holds;
(d) if $y \in \bar{\mathfrak{A}} - \mathfrak{A}$, then $\lim_{x \to y} v(x) = -1$, furthermore, $\lim_{|x| \to \infty} v(x) = -1$, provided that this limit process can be carried out for $x \in \mathfrak{A}$;

(e) $\left(\dfrac{dv\,(p(t, x, 0))}{dt}\right)_{t=0} \equiv \sum_{i=1}^{n} \dfrac{\partial v}{\partial x_i} f_i(x) = \varphi(x)(1 + v(x))\sqrt{1 + |f|^2}$ (21.2)

Corollary 1: If $f \in C_r$, the function $\varphi(x)$ can always be chosen such that v belongs to the class C_r in $\mathfrak{A}$.

Corollary 2: The theorem can also be expressed in terms of a positive function v, in order to be in agreement with the theorems of Sec. 4.

The auxiliary function $\varphi(x)$ is appreciably arbitrary. We shall attempt to determine it such that differential equation (21.2) can be solved in the most convenient manner. The square root occurring in this differential equation is a consequence of the change of scale (20.5). If t is replaced by s in accordance with (20.5), then (21.2) becomes

$$dv/ds = (1 + v)\varphi \qquad (21.3)$$

Because of the continuity of v, this function is defined in $\bar{\mathfrak{A}}$. From property (d) it follows that the equation $v(x) + 1 = 0$ determines the precise boundary of the stability domain which consists of phase trajectories according to Erugin [2]. If $1 + v > 0$ for all values of x, the equilibrium is asymptotically stable in the whole; this condition is necessary and sufficient.

Example: The partial differential equation (21.2) corresponding to

$$\dot{x} = -x + 2x^2 y, \qquad \dot{y} = -y \qquad (21.4)$$

is

$$\dfrac{\partial v}{\partial x}(2x^2 y - x) + \dfrac{\partial v}{\partial y}(-y) = \varphi(x, y)(1 + v)\sqrt{1 + y^2 + (2x^2 y - x)^2}$$

We put

$$\varphi(x, y) = \dfrac{x^2 + y^2}{\sqrt{1 + y^2 + (2x^2 y - x)^2}}$$

and obtain a differential equation with the solution

$$v = -1 + \exp\left(-\frac{y^2}{2} - \frac{x^2}{2(1-xy)}\right)$$

which vanishes at the origin and which is a Liapunov function for (21.4). From the condition $v + 1 = 0$, we obtain the curve $xy = 1$ as the boundary of stability.

Frequently, it will not be possible to obtain a closed form solution of the partial differential equation in spite of the freedom in the choice of $\varphi(x)$. In such cases, the precise determination of the stability boundary is not possible; however, it is often possible to obtain at least an estimate for $\mathfrak{A}$. Suppose the right hand side of (21.1) is analytic in the x_i, and let

$$\dot{x} = Ax + g(x) \tag{21.5}$$

where A is constant. The functions $g_i(x)$ have power series expansions in the neighborhood of the origin beginning with terms of at least second order. The real parts of the eigenvalues of A shall be negative; thus, the equation of the first approximation is stable (cf. Theorem 10.1). If here $\varphi(x)$ is chosen as a positive definite quadratic form and if the partial differential equation is solved for v, the series

$$v = v_2 + v_3 + \cdots \tag{21.6}$$

results, where v_j denotes a form of order j in the components of x. Since v is negative definite, the same holds for v_2 and, in particular, it is found that the derivative of v_2 for the linearized differential equation

$$\dot{x} = Ax$$

is equal to $\varphi(x)$.

Now let α be the least upper bound of the values of the function v_2 on the set defined by

$$(\dot{v}_2)_{(21.5)} = 0 \tag{21.7}$$

(with the exception of the origin). If such a set does exist, then $\alpha < 0$. This leads to:

THEOREM 21.2 (Zubov [3, 6]): The point set defined by

$$v_2(x) = \alpha \tag{21.8}$$

lies entirely within the domain of attraction $\mathfrak{A}$ of the equilibrium of (21.5).

Proof: If a subset of the set defined by (21.8) would lie outside $\mathfrak{A}$, then there would exist phase curves which would intersect the set (21.8) at two points; namely, those phase curves which, on the basis of a theorem of Erugin [2], form the stability boundary. Along that segment of the phase curves lying between both intersections, there would be a point $\bar{x}$ at which $\dot{v}_2$ would vanish since v_2 has the same value α at the end points of that segment. However, since $\bar{x}$ lies in the interior of the surface, or curve, respectively, defined by (21.8), we would have $v_2(\bar{x}) < \alpha$, and this would contradict the definition of α. The reasoning remains valid if there are points on the phase curve to which an equilibrium of (21.5) corresponds, since $\dot{v}_2$ vanishes in this case.

Analogously, one proves:

THEOREM 21.3 (Zubov [3, 6]): Let β be the greatest lower bound of the values of v_2 on the set (21.7). Then the domain of attraction lies entirely within the domain $v_2 > \beta$. If $\beta = -\infty$, the equilibrium is asymptotically stable in the whole.

From Theorems 21.2 and 21.3 it follows that the stability boundary lies completely in the domain

$$\alpha > v_2(x) > \beta$$

The domain of attraction for a given differential equation can be determined by means of Theorem 21.1. Zubov [3] showed that the converse problem can also be solved. Under certain assumptions, a differential equation can be established for a given domain $\mathfrak{A}$ such that $\mathfrak{A}$ is precisely the domain of attraction of the equilibrium.

Theorem 21.1 may also be formulated for an invariant set $\mathfrak{M}$ of a dynamical system in the phase space. Let $\mathfrak{A}$ be an open invariant set containing a neighborhood of the closed invariant set $\mathfrak{M}$, and let $\bar{\mathfrak{A}}$ be its closure.

THEOREM 21.4 (Zubov [6]): Necessary and sufficient that $\mathfrak{A}$ is precisely the stability domain of $\mathfrak{M}$ is the existence of two functions $v(x)$ and $\varphi(x)$ with the following properties:

(a) $v(x)$ is defined and continuous in $\mathfrak{A}$; $\varphi(x)$ is defined and continuous in the entire space;

(b) $-1 < v(x) < 0$ if $x \in \mathfrak{A}$, $x \notin \mathfrak{M}$; if $\rho(x, \mathfrak{M}) \neq 0$, then $\varphi(x) > 0$; if $x \in \mathfrak{M}$, then $\varphi(x) = 0$;

(c) for every sufficiently small number $c_1 > 0$, two numbers $c_2 > 0$ and $c_2' > 0$ can be found such that $v(x) < -c_2$ and $\varphi(x) > c_2'$ if $\rho(x, \mathfrak{M}) \geq c_1$;

(d) the functions v and φ tend towards zero as $\rho(x, \mathfrak{M}) \to 0$;

(e) $\lim v(x) = -1$ $(\rho(x, \bar{x}) \to 0)$ if a point $\bar{x} \notin \mathfrak{M}$, $\bar{x} \in \bar{\mathfrak{A}} - \mathfrak{A}$ exists;

(f) $\left(\dfrac{dv}{dt}\right)_{t=0} = \varphi(x)(1 + v(x))$.

The assumptions guarantee, incidentally, that $\mathfrak{M}$ is uniformly asymptotically stable and uniformly attractive (cf. Sec. 17, 6).

Theorem 21.4 permits statements to be made on the stability domain of nonautonomous differential equations with uniformly asymptotically stable equilibrium. The stability domain $\mathfrak{A}$ then is a cylindrical neighborhood of the t-axis. In this case, the Liapunov function v as well as the auxiliary function φ depend in general explicitly on t. The properties of being "definite" and of being "decrescent," required only for $t \geqq t_0 \geqq 0$ in the theorems of the direct method, must in this case be extended to $t \geqq t_0 > -\infty$, since otherwise the theory of dynamical systems could not be applied.

5 LIAPUNOV FUNCTIONS WITH CERTAIN PROPERTIES OF RATE OF CHANGE

22. ORDER NUMBER AND EXPONENTIAL STABILITY

(a) In the cases of asymptotic stability and uniform asymptotic stability, the solutions tend towards zero with increasing t. Nothing was said about the rate of decay of these solutions and unless further assumptions are made, nothing can be said about the asymptotic behavior of the Liapunov functions, the existence of which is guaranteed in the case of uniform asymptotic stability. Such assumptions refer either to the behavior of the solutions or to the form of the differential equations. So far, the following cases have received closer consideration:

(α) The comparison function of the inequality (17.7) has the special form

$$\beta|x_0| \exp(-\alpha(t - t_0)) \qquad (\alpha > 0, \beta > 0)$$

Then the existence of a positive definite Liapunov function v can be inferred which satisfies estimates of the form

$$v(x, t) < a|x|^\gamma, \quad |\dot{v}| > b|x|^\gamma \qquad (a > 0, b > 0, \gamma > 0)$$

This statement can be reversed. If the solutions increase like $\exp(\alpha(t - t_0))$, the order of magnitude of the Liapunov function can also be determined.

(β) For sufficiently large $t - t_0$, the comparison function has the form

$$c\sqrt{|x_0|}(t - t_0)^{-\beta} \qquad (c > 0, \beta > 0)$$

In this case there exists a function v with $|v| < a|x|^\gamma$, $|\dot{v}| > b|x|^{\gamma'}(\gamma' > \gamma)$. Conversely, from the existence of such a function it can be inferred that the solutions increase or decrease at least like a power of $t - t_0$.

(γ) The right hand sides of the differential equations are homogeneous linear functions with bounded coefficients. The treatment of this case is in particular necessary for stability investigations by means of the method of the first approximation for nonautonomous differential equations, which will be dealt with in Chap. 6. For the following, we need the concept of the *order number* of a function. This number can be determined by comparing the function under consideration with an exponential function.

DEFINITION 22.1: Let the real or complex valued function $f(t)$ of the real variable t be defined for $t \geq t_0$ and bounded for finite values of t. By the *order number* of the function $f(t)$ we shall understand the expression

$$\pi(f) = \limsup_{t \to \infty} \frac{\log |f(t)|}{t} \tag{22.1}$$

In most publications following Liapunov, the *characteristic number* of a function is used. The characteristic number of $f(t)$ is equal to $-\pi(f)$. The preceding definition is due to Perron [2].

The properties of the order number follow immediately from its definition:

$$\pi(f+g) = \max(\pi(f), \pi(g)) \quad \text{if} \quad \pi(f) \neq \pi(g)$$
$$\pi(f+g) \leq \pi(f) \quad \text{if} \quad \pi(f) = \pi(g) \tag{22.2}$$

$$\pi(fg) \leq \pi(f) + \pi(g) \tag{22.3}$$

$$\pi\left(\int_{t_0}^{t} f(u)\,du\right) \leq \pi(f) \quad \text{if} \quad \pi(f) \geq 0$$
$$\pi\left(\int_{t}^{\infty} f(u)\,du\right) \leq \pi(f) \quad \text{if} \quad \pi(f) < 0 \tag{22.4}$$

DEFINITION 22.2: By the order number of the vector $x(t)$ we understand the order number of its absolute value

$$\pi(x) = \limsup_{t \to \infty} \frac{1}{t} \log \sqrt{\sum_{i=1}^{n} x_i^2(t)} \tag{22.5}$$

Apparently,

$$\pi(x) = \limsup_{t \to \infty} \frac{1}{t} \log \max_{i} |x_i|$$
$$= \max(\pi(x_1), \cdots, \pi(x_n)) \tag{22.6}$$

The properties (22.2) to (22.4) are correspondingly valid for the expression (22.5).

(b) Now let

$$\dot{x} = f(x, t) \tag{22.7}$$

If all solutions with sufficiently small initial values have negative order numbers, then the equilibrium is certainly asymptotically stable, but not necessarily uniformly asymptotically stable. Both properties are combined in

DEFINITION 22.3: The equilibrium of the differential equation (22.7) is called *exponentially stable* if there exist two positive constants α and β which are independent of the initial values, such that for sufficiently small initial values the inequality

$$|\mathsf{p}(t, x_0, t_0)| < \beta |x_0| \exp(-\alpha(t - t_0)) \qquad (22.8)$$

is satisfied.

(Massera [4] speaks of exponential asymptotic stability. Following K. P. Persidskii [6], Krasovskii [13, 16] speaks of uniform asymptotic stability according to the first approximation.) In order to define more precisely the behavior of the solutions of a differential equation with unstable equilibrium, it is convenient to introduce two more concepts.

DEFINITION 22.4: The equilibrium of the differential equation (22.7) is called *exponentially unstable* if two positive constants α and β exist and if there are initial values (x_0, t_0) in every domain $\Re_{h,\tau}$ with arbitrarily small h and arbitrarily large τ such that

$$|\mathsf{p}(t, x_0, t_0)| > \beta |x_0| \exp(\alpha(t - t_0))$$

If this relation is valid for all initial points x_0 with sufficiently small absolute values, then the equilibrium is called *completely exponentially unstable*.

DEFINITION 22.5: The motions of differential equation (22.7) have *intensive behavior*, if every motion admits along at least one of its branches an estimate of the form

$$|\mathsf{p}(t, x_0, t_0)| > \beta |x_0| \exp(\alpha |t - t_0|)$$
$$(t \geq t_0, \text{ or } t \leq t_0) \qquad (22.9)$$

where $\alpha > 0$, $\beta > 0$. Exponential stability and complete instability are special cases of this concept which was introduced by Krasovskii [13] in a somewhat different terminology. In the cases of exponential stability, or exponential instability, we shall speak of *significant behavior* of the motions. The so-defined concept coincides with that introduced in Sec. 8 for the case of linear differential equations with constant coefficients. It should be emphasized that the existence of an estimate of the type

$$|\mathsf{p}(t, x_0, t_0)| = O(e^{-\alpha t}) \qquad (t \to \infty)$$

guarantees exponential decay of the solution and thereby asymptotic stability of the equilibrium, but not uniformity of stability and not at all exponential stability.

A condition for intensive behavior is expressed in:

THEOREM 22.1: (Krasovskii [13, 16]): *The existence of a Liapunov function $v(x, t)$ satisfying the estimates*

$$v(x, t) < a_1|x|^\gamma \qquad (22.10)$$

$$|\dot{v}| > a_2|x|^\gamma \qquad (22.11)$$

where a_1, a_2, and γ are positive constants, is necessary and sufficient for the motions of differential equation (22.7) to have intensive behavior.

Corollary: In the case of intensive behavior, $v \in C_1$ can be chosen such that in addition to (22.10) and (22.11) the inequalities

$$\left|\frac{\partial v}{\partial x_i}\right| < a_3|x|^{\gamma-1} \qquad (i = 1, \cdots, n; a_3 > 0, \gamma > 1) \qquad (22.12)$$

hold.

Proof:

(α) Assume inequalities (22.10) and (22.11) to be satisfied and, in particular, assume that initial values x_0, t_0 exist such that $v(x_0, t_0) > 0$, $\dot{v}(x_0, t_0) \geq 0$. According to Sec. 1(i), we introduce the function

$$v(t) = v(\mathsf{p}(t, x_0, t_0), t)$$

which increases with t since, as a consequence of (22.10) and (22.11),

$$\dot{v}(t) > a_4 v(t) \qquad (a_4 = a_2/a_1)$$

and hence

$$v(t) > v(t_0) \exp(a_4(t - t_0)) \qquad (22.13)$$

Consequently, a positive τ can be chosen such that for $t_1 = t_0 + \tau$ the inequality

$$v(t_1) > a_5|x_0|^\gamma$$

holds, where the constant a_5 is independent of x_0. Then it follows from (22.13) that for $t > t_1$

$$\frac{\dot{v}(t)}{v(t)} > \frac{a_2}{a_1} = a_4$$

$$v(t) > v(t_1) \exp(a_4(t - t_1))$$
$$= v(t_1) \exp(a_4(t - t_0)) \exp(-a_4\tau)$$

and since

$$a_1|\mathsf{p}(t, x_0, t_0)|^\gamma > v(t)$$

we finally find that

$$a_1|\mathsf{p}(t, x_0, t_0)|^\gamma > a_5 \exp(-\tau a_4)|x_0|^\gamma \exp(a_4(t - t_0)) \qquad (t > t_1)$$

This is an inequality of the form (22.9).

(β) If $v(t_0)$ and $\dot{v}(t_0)$ are of opposite sign in the preceding computations, the considerations have to be carried out for the time interval $t < t_0$. This again leads to an inequality of the form (22.9).

(γ) The following theorem is used to prove the necessity of the conditions of Theorem 22.1:

THEOREM 22.2: Let $w(\mathsf{x}, t) \in C_1$ be defined on a domain of the motion space and let

$$v(\mathsf{x}, t) = \int_t^{t+h} w(\mathsf{p}(\tau, \mathsf{x}, t), \tau)\, d\tau \qquad (22.14)$$

Along the path of integration, p shall remain in the domain of definition of w. Then the derivative for (22.7) is

$$\dot{v} = w(\mathsf{p}(t + h, \mathsf{x}, t), t + h) - w(\mathsf{x}, t) \qquad (22.15)$$

Proof: Forming the derivative $\dot{v}$ for (22.7), an expression is obtained which contains two integral terms in addition to the right hand side of (22.15); the integrand is

$$\sum_{i=1}^n \frac{\partial w}{\partial p_i} \dot{p}_i + \frac{\partial w}{\partial t} = \sum_{i,k} \frac{\partial w}{\partial p_k} \frac{\partial p_k}{\partial x_i} \dot{x}_i + \sum_{k=1}^n \frac{\partial w}{\partial p_k} \frac{\partial p_k}{\partial t}$$

$$= \sum_{k=1}^n \frac{\partial w}{\partial p_k} \left(\sum_{i=1}^n \frac{\partial p_k}{\partial x_i} f_i + \frac{\partial p_k}{\partial t} \right)$$

The expression in parentheses vanishes identically for all $k = 1, \ldots, n$ (cf., e.g., Kamke [1], Sec. 18, No. 87). Thus, only (22.15) remains.

If Theorem 22.2 is applied with $w = |\mathsf{x}|^2$, a function $v(\mathsf{x}, t)$ is obtained which satisfies the inequalities (22.10) to (22.12) for $\gamma = 2$. If the equilibrium is exponentially stable, h may tend towards ∞. The function obtained is positive according to the construction; it is even positive definite, which can be shown as follows: Let $|\mathsf{p}|^2 = \xi(\tau)$, then

$$\frac{d\xi}{d\tau} = 2 \sum_{i=1}^n p_i \frac{dp_i}{d\tau} = 2 \sum_{i=1}^n p_i f_i$$

and since $f(\mathsf{x}, t)$ is bounded in $\Re_{h,t_0}$, it follows that

$$|\xi'| = \left|\frac{d\xi}{d\tau}\right| < c|\mathsf{p}| = c\sqrt{\xi} \qquad (c > 0)$$

Consequently,

$$v(\mathsf{x}, t) = \int_{\tau=t}^{\infty} |\mathsf{p}(\tau, \mathsf{x}, t)|^2\, d\tau$$

$$= \int_{\xi=|\mathsf{x}|^2}^{\xi=0} \frac{\xi}{\xi'}\, d\xi$$

$$\geq \frac{1}{c} \int_0^{|\mathsf{x}|^2} \sqrt{\xi}\, d\xi > c_1 |\mathsf{x}|^3 \qquad (c_1 > 0)$$

i.e., v is positive definite. By (22.8) we obtain

$$v(\mathsf{x}, t) < |\mathsf{x}|^2 \beta^2 \int_t^\infty e^{-\alpha(\tau-t)}\, d\tau < c_2|\mathsf{x}|^2$$

which shows that v is also decrescent.

If the equilibrium is unstable, h is chosen such that $h > (1/\alpha) \log (1/\beta)$. Without verification, Krasovskii [16] supplied the function

$$v(\mathsf{x}, t) = \int_{t+h}^t |\mathsf{p}(\tau, \mathsf{x}, t)|^2\, d\tau + \int_{t-h}^t |\mathsf{p}(\tau, \mathsf{x}, t)|^2\, d\tau,$$

$$h = (4/\alpha) \log (2/\beta)$$

which is applicable in both cases and which increases faster along one branch of the motion than it decreases along the other. Krasovskii [13] developed a considerably more cumbersome topological construction of the function $v(\mathsf{x}, t)$ of Theorem 22.1.

If, in addition to (22.10) and (22.11), inequalities of the form

$$v > a_1'|\mathsf{x}|^\gamma, \qquad |\dot{v}| < a_2'|\mathsf{x}|^\gamma \qquad (22.14)$$

are valid, then the motion can be estimated above and below. For example, as in the proof of Theorem 22.1 and in analogy to (22.13), the inequalities

$$\frac{\dot{v}}{v} < \frac{a_2'}{a_1'} = a_4', \qquad v < v(t_1) \exp\left(a_4'(t - t_1)\right)$$

are obtained for $t > t_1 = t_0 + \tau$, and then as above

$$a_1'|\mathsf{p}(t, \mathsf{x}_0, t_0)|^\gamma < a_5'|\mathsf{x}_0|^\gamma \exp(-a_4'\tau) \exp(a_4'(t - t_0))$$

The considerations of part (β) of the proof of Theorem 22.1 can be transferred similarly. Thus we obtain the following extension of Theorem 22.1.

THEOREM 22.3: *The existence of a function v satisfying, for sufficiently small $|\mathsf{x}|$, inequalities of the form*

$$a_1'|\mathsf{x}|^\gamma < v(\mathsf{x}, t) < a_1|\mathsf{x}|^\gamma$$

$$a_2|\mathsf{x}|^\gamma < |\dot{v}(\mathsf{x}, t)| < a_2'|\mathsf{x}|^\gamma$$

is necessary and sufficient for every motion with sufficiently small values of $|\mathsf{x}_0|$ to admit an estimate of the form

$$\beta_1|\mathsf{x}_0| \exp(\alpha_1|t - t_0|) < |\mathsf{p}(t, \mathsf{x}_0, t_0)| < \beta_2|\mathsf{x}_0| \exp(\alpha_2|t - t_0|),$$

$$0 < \alpha_1 \leqq \alpha_2$$

along at least one of its branches.

23. DIFFERENTIAL EQUATIONS WITH HOMOGENEOUS RIGHT HAND SIDES

For the most part, the results of Sec. 22 can be extended to the case where the comparison function $\vartheta(r)$ of Theorem 17.4 is a power of r. The most important class of such differential equations consists of the equations

SEC. 23 LIAPUNOV FUNCTIONS WITH PROPERTIES OF CHANGE

with homogeneous right hand sides. The simplest equation of this type is

$$\dot{x} = -ax^k \qquad (a > 0, k \geq 2, k \text{ integer}) \qquad (23.1)$$

Here,

$$p(t, x_0, t_0) = (x_0^{1-k} + a(k-1)(t-t_0))^{-1/(k-1)} \qquad (23.2)$$

If k is an odd number, the equilibrium is uniformly asymptotically stable and, in particular, for $t - t_0 > 0$ the inequality

$$|p(t, x_0, t_0)| < |x_0|^{1/2}[a(k-1)(t-t_0)]^{-1/2(k-1)} \qquad (23.3)$$

is satisfied. Consequently, the functions

$$\kappa(r) = r^{1/2}$$

$$\vartheta(r) = (a(k-1)r)^{-1/2(k-1)}$$

can be chosen as comparison functions in the sense of Theorem 17.5. With suitably chosen constants, Eq. (23.2) can be written in the form

$$p^{1-k} = \beta x_0^{1-k} + \alpha(t - t_0)$$

This leads to:

DEFINITION 23.1: The motions of differential equation (2.7) have *weakly intensive behavior* if every motion with sufficiently small initial values admits, along at least one of its branches, an estimate of the form

$$|\mathsf{p}(t, x_0, t_0)|^{-\eta} < \beta|x_0|^{-\eta} - \alpha|t - t_0|$$
$$(t > t_0 \text{ or } t < t_0) \qquad (23.4)$$

Here, α, β, and η are positive constants.

In the case of asymptotic stability, relation (23.4) is valid for $t < t_0$. If t and t_0 are interchanged, it follows that

$$|\mathsf{p}(t, x_0, t_0)|^{-\eta} > \beta^{-1}|x_0|^{-\eta} + \alpha\beta^{-1}(t - t_0) \qquad (t > t_0) \qquad (23.5)$$

and for sufficiently large $t - t_0$, for sufficiently small $|x_0|$ and for suitable $\gamma > 0$, the inequality

$$|\mathsf{p}(t, x_0, t_0)| < \gamma\sqrt{|x_0|}(t - t_0)^{-1/2\eta} \qquad (23.6)$$

holds. In analogy to Theorem 22.1, the following theorem is obtained.

THEOREM 23.1: The existence of a function $v(x, t)$ satisfying inequalities of the form

$$v < a_1|x|^\gamma, \qquad |\dot{v}| > a_2|x|^{\gamma+\eta} \qquad (23.7)$$

for sufficiently small $|x|$ is necessary and sufficient for the motions of differential equation (2.7) to have weakly intensive behavior. Here, a_1, a_2, η and γ are positive constants.

Corollary: In the case of weakly intensive behavior, $v \in C_1$ can be determined such that, in addition to (23.7), the inequality

$$\left|\frac{\partial v}{\partial x_i}\right| < a_3 |x|^{\gamma-1} \qquad (i = 1, \ldots, n; a_3 > 0, \gamma > 1) \qquad (23.8)$$

is valid.

Proof: The reasoning is the same as in the proof of Theorem 22.1. In place of (22.13), we obtain the inequality

$$\frac{\dot v}{v^{1+\eta/\gamma}} > \frac{a_2}{a_1^{\gamma+\eta}} = a_4$$

and by integration

$$v^{-\eta/\gamma} - v(0)^{-\eta/\gamma} < -\frac{\eta}{\gamma} a_4 (t - t_0)$$

However, since

$$v^{-\eta/\gamma} > a_1^{-\eta/\gamma} |x|^{-\eta}$$

an inequality of the form (23.4) results.

Theorem 22.2 is needed to show the necessity of the condition. According to (23.5), the inequality

$$|\mathsf{p}(\tau, x, t)| \leq |x|(\beta + \alpha(\tau - t)|x|^\eta)^{-1/\eta}$$

is certainly valid in the case of stability. If a positive number is chosen for γ, the integral

$$v(x, t) = \int_t^\infty |\mathsf{p}(\tau, x, t)|^{\gamma+\eta}\, d\tau \qquad (23.9)$$

converges; it is of the type (22.14) with $w(x, t) = |x|^{\gamma+\eta}$. Now let

$$|x|^\eta (\tau - t) = u$$

then for $\tau \geq t$:

$$v \leq |x|^\gamma \int_0^\infty \frac{du}{(\beta + \alpha u)^{1+\gamma/\eta}}$$

This is the first inequality (23.7); it proves the decrescentness of v. The definiteness follows in the same way as in the proof of Theorem 22.1; Theorem 22.2 yields the equation

$$\dot v = -|x|^{\gamma+\eta}$$

In the case of an unstable equilibrium, an integral extended over a suitable finite interval is used.

Let the components $h_i(x_1, \ldots, x_n)$ of $\mathsf{h}(x)$ in the autonomous differential equation

$$\dot x = \mathsf{h}(x) \qquad (23.10)$$

be homogeneous functions of rational order $r > 0$. Then

$$c\mathsf{p}(c^{r-1}t, x_0, 0) = \mathsf{p}(t, cx_0, 0) \qquad (23.11)$$

Thus, the equilibrium is asymptotically stable in the whole if it is asymptotically stable.

THEOREM 23.2 (Krasovskii [13]): Let the order $m > 1$ of the functions h_i be an integer. The differential equation (23.10) shall not have a bounded solution for $-\infty < t < +\infty$ other than the trivial solution. Then there exists a Liapunov function $v \in C_1$ which is independent of t and which satisfies the estimate

$$|\dot{v}| > a_2|\mathsf{x}|^{\gamma+m-1} \tag{23.12}$$

in addition to the inequalities (22.10) and (22.12).

Krasovskii [13] provided a construction procedure for these functions which is based on the same considerations as his proof of Theorem 19.1. By means of Theorem 23.1, weakly intensive behavior is inferred from the existence of these functions, and it is recognized that a Liapunov function can be constructed under the assumption that v is of the form (23.9). In addition, if relation (23.11) is used, we see that the functions v and $\dot{v}$ can be determined as homogeneous functions of the coordinates $x_1, \ldots, x_n$.

Zubov [11] proved the following stability criterion for differential equations of the form (23.10): The existence of a Liapunov function $v(\mathsf{x})$ with the following properties is necessary and sufficient for asymptotic stability:

1. The total derivative of $v(\mathsf{x})$ for (23.10) is a negative definite form of the order $k > r + 1$;

2. The form $v(\mathsf{x})$ and its partials satisfy the equation

$$\sum_{i=1}^{n} \frac{\partial v}{\partial x_i} \dot{x}_i = (k + 1 - r)v$$

From Theorems 23.1 and 23.2 follows:

THEOREM 23.3 (Krasovskii [13]): If the autonomous differential equation (23.10), the order of which is supposed to be an integer, does not have a bounded solution for $-\infty < t < +\infty$, then all of its motions have weakly intensive behavior.

It is not yet known whether Theorems 23.2 and 23.3 remain valid if the functions $h_i(\mathsf{x})$ are homogeneous, but of nonintegral order. However, the following theorem holds:

THEOREM 23.4 (Zubov [6]): Let the components h_i of the right hand side of the autonomous differential equation (23.10) be homogeneous functions of positive rational order $\mu \neq 1$. Estimates of the form (23.6) are assumed to be known for the solutions. Then estimates of the form

$$c_1'|\mathsf{x}_0|^{1-\mu} + c_2'(t - t_0) < |\mathsf{p}(t, \mathsf{x}_0, t_0)|^{1-\mu} < c_1''|\mathsf{x}_0|^{1-\mu} + c_2''(t - t_0)$$

are valid. (In the case $\mu > 1$, the number μ must be of the form $(2p + 1)/q$, where p and q are integers). All motions have the same order of magnitude with respect to $t - t_0$.

Conjecture: Theorem 23.3 is very likely also valid for differential equations whose right hand sides h_i are homogeneous functions of nonintegral order. For the case of nonautonomous equations with homogeneous right hand sides, the following theorem seems to be true: If the equilibrium of such a differential equation is uniformly asymptotically stable, then its motions satisfy an estimate of the form (23.6).

By means of his own method (cf. Sec. 21), Zubov showed that inequality (23.6) may be replaced by a quite different statement: The following condition is necessary and sufficient for every solution of (23.10) to satisfy an estimate of the form (23.5): The domain of attraction of the system of differential equations

$$\dot{x}_i = -x_i - h_i(\mathsf{x}) \qquad (i = 1, \ldots, n)$$

is bounded.

This domain of attraction can be investigated directly (cf. Sec. 21). Therefore, it is sometimes possible to decide whether or not (23.6) is satisfied.

24. THE STABILITY BEHAVIOR OF LINEAR DIFFERENTIAL EQUATIONS

The following theorem is a consequence of the results of the remark to Theorem 8.2. The terminology of Sec. 22 is used.

THEOREM 24.1: *The motions of a linear differential equation with constant coefficients have significant behavior if and only if either all roots of the characteristic equation have negative real parts, or if there is at least one root with positive real part. In the first case the equilibrium is exponentially stable, in the second case it is exponentially unstable.*

An equivalent statement is valid for periodic linear differential equations

$$\dot{\mathsf{x}} = \mathsf{A}(t)\mathsf{x} \qquad (\mathsf{A}(t+w) = \mathsf{A}(t)) \tag{24.1}$$

with continuous and bounded coefficients.

THEOREM 24.2: *The motions of a linear differential equation with periodic coefficients have significant behavior if and only if its characteristic equation has either only roots with absolute value less than unity, or there is at least one root with absolute value greater than unity. In the first case the equilibrium is exponentially stable, and in the second case it is exponentially unstable.*

In order to obtain the characteristic equation of (24.1) (its definition is different from that of the characteristic equation of autonomous linear differential equation, cf., e.g., Malkin [19]), the linear substitution S is considered to which the fundamental system $\mathsf{X}(t)$ of (24.1) is subjected if the argument t is replaced by $t + w$:

$$X(t + w) = X(t)S \qquad (24.2)$$

The characteristic equation is then the eigenvalue equation of the matrix S, i.e.,

$$\det(S - \lambda U) = 0 \qquad (24.3)$$

Equation (24.3) is independent of the particular choice of a fundamental system. The proof of Theorem 24.2 can be accomplished by means of the theory of difference equations (cf. Sec. 39), or by the so-called Reducibility Theorem of Liapunov [1].

DEFINITION 24.1 (Liapunov [1]): A linear differential equation with continuous and bounded coefficients is said to be *reducible*, if it can be transformed by means of a nonsingular linear substitution

$$y = Q(t)x \qquad (24.4)$$

with bounded coefficients into a linear differential equation with constant coefficients,

$$\dot{y} = Ry \qquad (24.5)$$

Obviously, the existence of a matrix $Q(t)$ which satisfies a differential equation of the form

$$\dot{Q}Q^I + QAQ^I = R, \qquad R = \text{const}$$

is necessary and sufficient for reducibility (cf. also Yakubovich [1]).

According to Liapunov, a periodic differential equation is reducible. In particular, a periodic matrix $Q(t)$ can be constructed by means of a fundamental system of differential equation (24.1), or of the adjoint differential equation

$$\dot{x} + A^T(t)x = 0 \qquad (24.6)$$

respectively. The matrices R and S are connected by the relation

$$R = (1/w) \log S$$

The equilibrium of (24.5) has the same stability property as the equilibrium of (24.1). Thus, Theorem 24.2 follows by applying Theorem 24.1 to Eq. (24.5).

In the case of autonomous, or periodic differential equations, uniform asymptotic stability is (by Theorem 17.5) a consequence of asymptotic stability. Using Theorems 24.1 and 24.2 even exponential stability can be inferred from the asymptotic stability in the case of linear autonomous, or periodic differential equations. However, this conclusion is in general not applicable for nonautonomous differential equations, as may be shown by the equation mentioned in Sec. 17,

$$\dot{x} = -\frac{x}{1+t} \qquad (24.7)$$

whose solution

$$x(t) = \frac{x_0}{1+t}(1+t_0)$$

does not tend exponentially towards zero. However, the following theorem is valid:

THEOREM 24.3: If the equilibrium of a linear differential equation is uniformly asymptotically stable, then it is also exponentially stable.

Proof: Let x_{0i} ($i = 1, \ldots, n$) be n linearly independent initial vectors; let $\mathsf{p}(t, x_{0i}, t_0)$ be the corresponding linearly independent solutions and let $\mathsf{X}(t, t_0)$ be the matrix of this fundamental system. Furthermore, let

$$\mathsf{Z}(t, t_0) = \mathsf{X}(t, 0)\mathsf{X}^I(t_0, 0)$$

Then the "general solution" is

$$\mathsf{p}(t, x_0, t_0) = \mathsf{Z}(t, t_0)x_0 \qquad (24.8)$$

and

$$|\mathsf{p}(t, x_0, t_0)|^2 = |\mathsf{Z}x_0|^2 = x_0^T \mathsf{Z}^T \mathsf{Z} x_0 \qquad (t \geq t_0)$$

Denoting the largest eigenvalue of the matrix $\mathsf{Z}^T\mathsf{Z}$, in other words, denoting the norm of the linear operator Z, by $\mu^2(t, t_0)$, we obtain the inequality

$$|\mathsf{p}(t, x_0, t_0)|^2 \leq \mu^2(t, t_0)|x_0|^2$$

Comparing this expression with inequality (17.7), which holds by assumption, we find, that $\kappa(r)$ can be chosen as $\kappa(r) = r$ and that $\mu(t, t_0)$ satisfies the inequality $\mu(t, t_0) \leq \vartheta(t - t_0)$. Now, for $t_0 < t_1 < t$, the relation

$$\mathsf{Z}(t, t_0) = \mathsf{Z}(t, t_1)\mathsf{Z}(t_1, t_0)$$

is valid, from which

$$\mu(t, t_0) \leq \mu(t, t_1)\mu(t_1, t_0)$$

follows. Repeated application of this inequality for $t = t_0 + n\tau$ ($n = 1, 2, \ldots$) yields the relation

$$\mu(t, t_0) \leq (\mu(t_0 + \tau, t_0))^n$$

If now τ is chosen such that $\vartheta(\tau) < \tfrac{1}{2}$ and if

$$\alpha = (\log 2)/\tau$$

then

$$\mu(t, t_0) < 2 \exp(-\alpha(t - t_0))$$

and since $|\mathsf{p}| < \mu(t, t_0)|x_0|$, it follows that the equilibrium is exponentially stable.

Theorem 24.3, formulated independently here, is a part of Malkin's [19, Sec. 75] proof of the following theorem:

THEOREM 24.4 (K. P. Persidskii [4]): If a decrescent and positive definite function v exists for the linear differential equation

$$\dot{x} = \mathsf{A}(t) x \qquad (24.10)$$

and if the total derivative $\dot v$ for (24.10) is negative definite, then the equilibrium is exponentially stable.

(Liaščenko [1] established another sufficient condition for exponential stability.) Theorem 24.4 follows from Theorems 18.3 and 24.3. Kuzmin [1] showed that in cases where $\dot v$ is not negative definite, stability, or asymptotic stability, respectively, can occasionally be inferred. Somewhat more precise statements can be made about the Liapunov function guaranteed by Theorem 22.1 if the equilibrium of (24.10) is exponentially stable. This is the content of

THEOREM 24.5 (Malkin [15]): If the equilibrium of (24.10) is exponentially stable, then a decrescent, positive definite function $v(x, t)$ can be constructed such that its derivative $\dot v$ is equal to an arbitrarily given negative definite form $-w(x, t)$ of integral order m and with continuous and bounded coefficients. Specifically,

$$v(x, t) = \int_t^\infty w(\rho(\tau, x, t), \tau)\, d\tau \qquad (24.11)$$

Proof: Let a_1, a_2 and a_3 be positive constants. Certainly,

$$a_1|x|^m \leq w(x, t) \leq a_2|x|^m$$

and from (22.8)

$$v(x, t) \leq a_3|x|^m$$

which means that $v(x, t)$ is decrescent. The definiteness can be inferred as in the proof of Theorem 22.1. The rest of the statement is obtained from Theorem 22.2.

As in the case of Theorem 22.1, a Liapunov function analogous to (24.11) can be found if the equilibrium of (24.10) is completely exponentially unstable. In this case, the integration has to be carried out over a suitable finite interval $t \leq \tau \leq t + h$.

Theorems 24.4 and 24.5 have also been found by Antosiewicz and Davis [1].

If w is chosen as a quadratic form, $w = x^T Q(t) x$, then v will also be a quadratic form, $v = x^T P(t) x$. Gorbunov [1 to 4] thoroughly investigated the relations between these two forms without imposing conditions on the stability behavior of (24.10). The relation between P, Q and A is

$$\dot P + A^T P + PA = -Q \qquad (24.12)$$

Denoting the minimum of the form w for fixed t under the side condition $v = 1$ by $\nu(t)$, we obtain the following

THEOREM 24.6 (Gorbunov [3]): A necessary and sufficient condition for the asymptotic stability of the equilibrium of (24.10) is: there exist two quadratic forms $v = x^T P x$, $w = x^T Q x$, related by (24.12), which are positive definite for every finite fixed t and of such a nature that the integral

$$\int_{t_0}^{t} \nu(\tau)\, d\tau$$

increases beyond any bound as $t \to \infty$.

Apparently, $\nu(t)$ is equal to the smallest eigenvalue of QP^I; therefore, this theorem belongs to the realm of Sec. 15.

We consider the system of equations

$$\dot{x}_i = \sum_{j=1}^{n} a_{ij}(t)\, x_j \qquad (i = 1, \ldots, n-1)$$

$$\mu \dot{x}_n = \sum_{j=1}^{n} a_{nj}(t)\, x_j$$

where μ is a small parameter. The question of interest is whether or not conclusions on the stability behavior of this system can be drawn from the stability behavior of the equilibrium of the system of order $n - 1$ which corresponds to the value $\mu = 0$. If the equilibrium of the degenerate system is exponentially stable, a quadratic form can be constructed as a corresponding Liapunov function. If the term

$$\mu x_n (g_1(t)\, x_1 + \cdots + g_n(t)\, x_n)$$

is added so that a Liapunov function for the complete system is obtained, and if the definiteness conditions are evaluated, the following statement results: If the absolute value of the coefficient $a_{nn}(t)$ is always larger than a fixed positive number, then exponential stability of the complete system follows from exponential stability of the degenerate system, provided μ is sufficiently small and is opposite in sign to $a_{nn}(t)$. (Razumichin [6], *without* the assumption of exponential stability. Cf. also Gradštein [1].)

If the equilibrium of (24.10) is stable, a quadratic form v can easily be found which satisfies the assumptions of Theorem 4.1. To compute this form, we may apply the method of K. P. Persidskii [4] which was mentioned in connection with Theorem 18.1. Because it follows from (24.8) that $x_0 = X(t_0, 0)\, X^I(t, 0)\, p$. Consequently,

$$v(x, t) = |X^I(t, 0) x|^2$$

is a Liapunov function of the desired kind. Gorbunov [2] determined such quadratic forms in a different manner.

25. THE ORDER NUMBERS OF A LINEAR DIFFERENTIAL EQUATION

(a) Every solution of the linear differential equation (24.10) has a well determined order number (cf. Definition 22.2). In particular, the following theorem is valid:

THEOREM 25.1 (Liapunov [1]): Every solution of (24.10) (except the trivial) has a *finite* order number.

Proof: Let
$$z = e^{\lambda t}x \qquad (25.1)$$
be a new variable, and let λ be a parameter. Then
$$\frac{d}{dt}|z|^2 = z^T S(t,\lambda)z \quad \text{with} \quad S(t,\lambda) = A^T + A + 2\lambda U \qquad (25.2)$$
Since $A(t)$ is bounded, two values λ_1 and λ_2 can be chosen such that the corresponding matrices S have only positive and negative eigenvalues, respectively, for $t \geq 0$. Then, with suitably determined constant $\alpha > 0$,
$$2\alpha|z|^2 < z^T S(t,\lambda_1)z, \qquad -2\alpha|z|^2 > z^T S(t,\lambda_2)z$$
for $t \geq 0$. By integration of (25.2), the inequalities
$$e^{\alpha t}|z_0| < |z| \quad (\lambda = \lambda_1) \quad \text{or} \quad e^{-\alpha t}|z_0| > |z| \quad (\lambda = \lambda_2)$$
follow, and from (25.1) it can be concluded that the order numbers of the solutions $x(t)$ of (24.10) lie in the interval $(\alpha - \lambda_1, \alpha - \lambda_2)$.

If the differential equation has the particular form (10.7), the same procedure as above shows that the largest order number is at most equal to the largest real part of the eigenvalues of A. Because, if the linear substitution (25.1) is carried out and if the quadratic form $v = z^T B z$ is chosen as a Liapunov function, where B is determined by
$$A^T B + BA = -(U + 2\lambda B)$$
(cf. Sec. 8), then
$$\dot{v} = -|z|^2 - z^T(Q^T B + BQ)z$$
As a consequence of the assumption about Q, $\dot{v}$ is negative definite. The equilibrium of the differential equation in z is then asymptotically stable if and only if the matrix B is positive definite; i.e., if and only if the matrix $A + \lambda U$ has only eigenvalues with negative real parts. Then the largest real part occurring is the least upper bound for the admissible λ-values.

In simple cases, the order numbers can be compared with the values of a suitable Liapunov function. By this, estimates of the order numbers can be obtained (cf. also Sec. 15). From the inequality for $|x_k|$ in Sec. 15, Gorbunov [4] immediately obtained upper bounds for the order numbers.

(b) Let $x_1, \ldots, x_n$ be any fundamental system of differential equation (24.10). Because of Theorem 25.1, the sum
$$\pi(x_1) + \cdots + \pi(x_n) \qquad (25.3)$$
is bounded; for certain fundamental systems it takes on its minimum. Thus, the following definition is meaningful:

DEFINITION 25.1 (Liapunov [1]): *A fundamental system for which the sum* (25.3) *takes on its minimum, is called normal;* the order numbers of a normal fundamental system (arranged according to their numerical values) are called the *order numbers of the differential equation*. (They are uniquely determined.)

If the order numbers of a fundamental system are all distinct, the fundamental system is already normal. Because of the property (22.2), the sum (25.3) can never become smaller in the transition to another fundamental system.

If $X(t)$ is the matrix of a normal fundamental system, if $s(t)$ is its trace and if $d(t)$ is its determinant, then (cf., e.g., Kamke [1], Sec. 19)

$$\pi(d) = \limsup_{t \to \infty} \frac{1}{t} \int_0^t s(u)\, du$$

since (25.4)

$$d(t) = \exp\left(\int_0^t s(u)\, du\right)$$

(The initial instant t_0 which was of importance in Sec. 24 is unimportant here; it may be put equal to zero.) If $\pi(d)$ is estimated in accordance with (22.2) and (22.3) by means of the order numbers $\lambda_1, \ldots, \lambda_n$ of the differential equation, and if (22.4) is considered, then

$$\pi(d) \leq \lambda_1 + \cdots + \lambda_n \tag{25.5}$$

DEFINITION 25.2 (Liapunov [1]): The differential equation (24.10) is called *regular* if
1. the sum of its order numbers equals the order number of $d(t)$, and if
2. the relation $\pi(d) + \pi(1/d) = 0$ holds.

It is readily concluded that every autonomous linear differential equation and, more generally, every reducible (Definition 24.1) differential equation is regular. The following theorem is valid for regular differential equations:

THEOREM 25.2 (Liapunov [1], Perron [2]): Let differential equation (24.10) be regular; let $\lambda_1 \leq \lambda_2 \leq \cdots \leq \lambda_n$ be its order numbers and let $\mu_1 \geq \mu_2 \geq \cdots \geq \mu_n$ be the order numbers of its adjoint differential equation

$$\dot{y} + A^T(t)\, y = 0 \tag{25.6}$$

Then

$$\lambda_r + \mu_r = 0 \quad (r = 1, \ldots, n) \tag{25.7}$$

Proof: If Y is the matrix of a normal fundamental system of (25.6) and if X has its former meaning, then $X^T Y = \text{const}$, as can easily be verified by differentiation. From (22.3) it follows that $\lambda_r + \mu_r \geq 0$. On the other hand, $Y = X^{1T} C$. Consequently, the elements of Y can be expressed in terms of the determinant $d(t)$ and its minors and the order numbers of the elements of Y can be estimated by means of (22.3) and (22.2). Taking the regularity into account, it is recognized that $\lambda_r \leqq -\mu_r$. This proves the truth of the statement of the theorem. At the same time, it is seen that differential equation (25.6) is regular. Relation (25.7) is, by the way, sufficient for the regularity of the differential equations (24.10) and (25.6) (Perron [2]).

6 THE SENSITIVITY OF THE STABILITY BEHAVIOR TO PERTURBATIONS

26. STABILITY ACCORDING TO THE FIRST APPROXIMATION

Let the stability behavior of the equilibrium of the differential equation

$$\dot{x} = f(x, t) \qquad (f \in C_1) \tag{26.1}$$

be known. In addition, let the modified differential equation

$$\dot{x} = f(x, t) + g(x, t) \qquad (f + g \in C_1) \tag{26.2}$$

be given. Then the problem of the *sensitivity* of the stability behavior of Eq. (26.1) arises; in other words, the question arises as to the effect of the additional term $g(x, t)$ on the stability or instability of the equilibrium. In one of the simplest cases where f is a linear function with constant coefficients, this question is partially answered by Theorem 10.1. It shows that the stability behavior for a sufficiently small additional term is insensitive, provided the trajectories have significant behavior. The requirement "sufficiently small" can be more closely described. On the other hand, the stability is sensitive in the case of critical behavior, for even an arbitrarily small additional term can change the stability behavior.

Theorem 10.1 can be considerably extended and generalized by application of the results of Chap. 5, but a general answer to the question above is not yet available.

Theorem 26.1 (Krasovskii [13]): *Let the motion of* (26.1) *have intensive behavior and for sufficiently small* $|x| < \delta$ *let there be an estimate of the form*

$$|g(x, t)| < b|x| \qquad (b > 0) \tag{26.3}$$

for the additional term of (26.2). Then, if b is sufficiently small, the differential equations (26.1) and (26.2) have the same stability behavior.

Proof: (Krasovskii [13]): A Liapunov function is constructed according to Theorem 22.1 and its total derivatives are formed for (26.1) and (26.2). Because of (22.11) and (22.12) (in the case where $\dot{v}_{(26.1)} < 0$) the inequality

$$\dot{v}_{(26.2)} = \dot{v}_{(26.1)} + \sum_{i=1}^{n} \frac{\partial v}{\partial x_i} g_i < -a_2|x|^\gamma + nba_3|x|^\gamma$$

holds. For sufficiently small b, the function $\dot{v}_{(26.2)}$ is also negative definite. Thus the same stability theorem can be applied to both differential equations. (A different proof for the case of stability was given by Barbašin and Skalkina [1].)

Conjecture: It seems that Theorem 26.1 is valid even with the assumption "significant" instead of "intensive."

The most important application of Theorem 26.1 is made with regard to the question of stability according to the first approximation. If (under the assumption $f \in C_\infty$)

$$\left.\frac{\partial f_i}{\partial x_k}\right|_{x=0} = a_{ik}(t), \quad A(t) = (a_{ik}(t))$$

holds, then Eq. (26.1) can be written in the form

$$\dot{x} = A(t)x + g(x, t) \qquad (26.4)$$

This is an equation of the type (26.2). The "reduced" differential equation with which it has to be compared is linear:

$$\dot{x} = A(t)x \qquad (26.5)$$

Since the series expansions of the components of $g(x, t)$ begin with terms of at least second order, an estimate of the form (26.5) can be made.

THEOREM 26.2: *If the motions of the linearized differential equation (26.5) have intensive behavior, then the complete and the reduced differential equation have the same stability behavior.*

Corollary: If the reduced differential equation is either autonomous, or periodic, then Theorem 26.2 is valid even with "significant" instead of "intensive." This follows from Theorem 10.1 in connection with Theorems 24.1 and 24.2. This corollary is often applied in practice if the stability of the periodic solution of an autonomous, or periodic differential equation is to be studied. For instance, let the canonical system (7.1)

$$\dot{q}_i = \frac{\partial H}{\partial p_i}, \quad \dot{p}_i = -\frac{\partial H}{\partial q_i} \quad (i = 1, \ldots, n)$$

be given, and let $\bar{p}_i, \bar{q}_i$ be the components of a special solution. In analogy to (2.5), the expressions

$$q_i = \bar{q}_i + x_i, \qquad p_i = \bar{p}_i + y_i$$

are substituted into the equations of motion and the right hand sides of these equations are expanded with respect to powers of the x_i and y_i. Then the differential equations of the perturbed motion are obtained (cf. Eq. (2.6)):

$$\dot{x}_i = \sum_{k=1}^{n} \left(\frac{\partial^2 H}{\partial p_i \partial q_k} x_k + \frac{\partial^2 H}{\partial p_i \partial p_k} y_k \right)^* + \varphi_i$$

$$\dot{y}_i = - \sum_{k=1}^{n} \left(\frac{\partial^2 H}{\partial q_i \partial q_k} x_k + \frac{\partial^2 H}{\partial q_i \partial p_k} y_k \right)^* + \psi_i$$

The asterisk indicates that the quantities p̄ and q̄ are to be substituted for p and q. The functions φ_i and ψ_i contain the terms of higher order. The linear part represents the "equations in the variations"; it is periodic if the special solution is periodic. Due to the particular form of the linear part, sometimes a Liapunov function can be given immediately. For example, if

$$v = x_1 y_1 + \cdots + x_n y_n$$

is used, it can be seen that the equilibrium is unstable if the quadratic forms which correspond to the matrices

$$\left(\frac{\partial^2 H}{\partial p_i \partial p_k} \right)^* \quad \text{and} \quad \left(- \frac{\partial^2 H}{\partial q_i \partial q_k} \right)^*$$

are both positive definite (Pozarickii [1]). This result is valid also for non-periodic p, q. (Pozarickii treated the problem in the case of somewhat more general equations of motion in another paper [2].) Cf. also Theorem 23.1.

> *Problem:* It is not yet known whether stability according to the first approximation also exists for differential equations with almost-periodic linear part. This problem is closely related to that formulated in Sec. 17.

A special case of a linear differential equation with an almost-periodic right hand side and an exponentially stable equilibrium was treated by Štelik [2].

For considerations such as the previous ones it is generally assumed that the right hand side of the differential equation of the perturbed motion is at least continuous. However, Aizerman and Gantmakher [1] showed recently that one can also allow finite jumps. For this purpose, the equation of the first approximation must be defined in a suitable manner.

Perron [4] presented the following criterion: The matrix $A(t)$ shall be of such a nature that the nonhomogeneous differential equation

$$\dot{x} = A(t)x + f(t)$$

where $f(t)$ is arbitrary, but continuous and bounded, has only bounded solutions. Then stability exists according to the first approximation. It can be shown (Malkin [19]) that Perron's condition is equivalent to exponential stability of the equilibrium of (26.1). (Cf. also Massera and Schäffer [1].)

The conditions of Theorems 26.1 and 26.2 are not necessary conditions. For instance, the requirement of exponential stability can be weakened.

THEOREM 26.3 (Malkin [4]): Let the estimate

$$|g(x, t)| < b|x|^r \qquad (r > 1) \tag{26.6}$$

be valid for the nonlinear part of (26.4), and for the solutions of the linearized differential equation (26.5) let (22.8) be replaced by

$$|p(t, x_0, t_0)| < \beta e^{\rho t_0} e^{-\alpha(t-t_0)} \qquad (t > t_0)$$

with $0 < \rho < (2r - 1)\alpha$. Then the equilibrium of the complete differential equation is asymptotically stable and we even have

$$\lim_{t \to \infty} e^{\gamma t} x(t) = 0$$

provided $\gamma < \rho/(2r - 1)$.

Similar theorems can be found in papers of K. P. Persidskii [2], and Reghiş [1].

Without excluding exponential stability, the assumptions can be changed by replacing (26.3) by

$$|g(x, t)| < \varphi(t) |x|$$

where $\varphi(t)$ is subjected to certain limitations in rate of change (Germaidze [1]). Corduneanu [2] established a criterion in which a condition of the form $|g| < w(|x|, t)$ appears. Here, the scalar function w is of such a nature that the equilibrium of a certain scalar differential equation

$$y' = py^\alpha + qw(y^\alpha, t)$$

is asymptotically stable. Somewhat differently oriented are some results on the *linearly perturbed* differential equation

$$\dot{x} = (A(t) + B(t)) x$$

in comparison with the unperturbed equation $\dot{x} = A(t) x$. The best result available was given by Massera [4].

In the case of differential equations with weakly intensive behavior as treated in Sec. 23, the solutions are not compared with exponential functions, but with powers of t. In this case it is also possible to formulate theorems on the stability behavior of perturbed differential equations.

THEOREM 26.4 (Krasovskii [13]): Let the motions of the differential equation (26.1) be weakly intensive and let the additional term of (26.2) satisfy an estimate of the form

$$|g(x, t)| < b|x|^\eta$$

104   THE SENSITIVITY OF THE STABILITY BEHAVIOR   CHAP. 6

where η is the number defined in (23.4) and where b is sufficiently small. Then the differential equations (26.1) and (26.2) have the same stability behavior.

The assumptions are satisfied, for instance, if (26.1) is autonomous with homogeneous right hand sides of integral order and if (26.1) does not have solutions bounded for all t (cf. Theorem 23.3).

This theorem can be proved in the same manner as Theorem 26.1 by using the Laipunov function whose existence is guaranteed by Theorem 23.1. At the same time, it follows from Theorem 18.1 that the equilibrium of the perturbed differential equation is uniformly asymptotically stable. This theorem was originally proved by Malkin [14] for the case of differential equations with autonomous principal part. By a different approach Massera [4] showed the uniformity of the asymptotic stability directly.

If the components of the right hand side of (26.1) are homogeneous polynomials and if the equilibrium is unstable, then instability of the equilibrium cannot be inferred in general, not even if (26.1) is autonomous, as Krasovskii [13] showed by a counter example. To infer instability, the existence of bounded solutions must be excluded.

A related criterion was given by Šestakov [1].

Two special theorems will be mentioned here, which partially belong to the realm of *total* stability (cf. Sec. 28).

THEOREM 26.6 (Krasovskii [8]): Let (26.1) be autonomous; let the equilibrium be unstable and let there be a neighborhood of the origin which does not contain any complete trajectory. Then there exists a continuous function

$$\eta(x) \qquad (\eta(x) > 0 \quad \text{for} \quad x \neq 0, \eta(0) = 0)$$

with the following property:
If

$$\sum_{i=1}^{n} g_i(x, t) < \eta(x)$$

the equilibrium of (26.2) is also unstable.

THEOREM 26.7 (Malkin [18]): Let the differential equation (26.1) be autonomous, or periodic, with an asymptotically stable equilibrium. In a certain neighborhood of the origin, $g(x, t)$ tends uniformly towards 0 with increasing t. Then the equilibrium of (26.2) is also asymptotically stable.

Remark 3 of Sec. 4 is used in the proof.

27. THE THEOREM OF LIAPUNOV ON REGULAR DIFFERENTIAL EQUATIONS

In a noteworthy theorem on stability according to the first approximation, the concept of the regular differential equation (cf. Definition 15.2) is used.

THEOREM 27.1 (Liapunov [1]): Let differential equation (26.4) be given and let the differential equation (26.5) of its first approximation be regular. If its order numbers are all negative, the equilibrium of (26.4) is asymptotically stable; if at least one of its order numbers is positive, the equilibrium of (26.4) is unstable.

Proof (According to Chetaev [7]; the original proof of Liapunov does not make use of the direct method): The notation of the proof of Theorem 25.2 shall be used. The fundamental systems X and Y shall be normalized such that $X^T Y = U$. Furthermore, let

$$\lambda_1 \leq \lambda_2 \leq \cdots \leq \lambda_n < \beta < 0$$

and let D be the diagonal matrix with the diagonal elements

$$e^{\lambda_1 t}, \ldots, e^{\lambda_n t}$$

If the new variable

$$z = D Y^T e^{\beta t} x$$

is introduced, then by Theorem 25.2 and by (22.2) and (22.3):

$$\pi(z) \leq \pi(x) + \beta$$

On the other hand, $x = e^{-\beta t} X D^I z$ and from this it follows that

$$\pi(x) \leq \pi(z) - \beta$$

i.e., $\pi(z) = \pi(x) + \beta$. Now, let $2v = |z|^2$. If the total derivative of v is formed for the differential equation in z obtained from (26.4), the relation

$$\dot{v} = z^T(\dot{D}D^I + \beta U) z + z^T(e^{\beta t} D Y^T g(x, t)) z \qquad (27.1)$$

follows. In the nonlinear part, all terms are of at least fourth order with respect to the components z_i of z. The matrix $e^{\beta t} D Y^T$ tends towards zero with increasing t; since, because of the regularity, the characteristic numbers of the variable vectors y_r are equal to $-\lambda_r$, and β is negative. Therefore, the second term $r(z, t)$ of (27.1) satisfies the estimate

$$|r(z, t)| < a|z|^2 \qquad (|z| \leq \delta, t \geq t_1, a > 0) \qquad (27.2)$$

with properly chosen numbers $\delta > 0$ and $t_1 > 0$. Since

$$z^T(\dot{D}D^I + \beta U) z = \sum_{i=1}^{n} (\lambda_i + \beta) z_i^2$$

it follows that

$$\dot{v} = \frac{1}{2} \frac{d}{dt} |z|^2 \leq (\lambda_n + \beta + a) |z|^2 \qquad (t \geq t_1)$$

If the initial point z_0 is chosen such that the corresponding solution $z(t)$ of the differential equation for z remains for $t_0 \leq t \leq t_1$ within the domain

defined by (27.2), then there exists a constant $c > 0$ such that for sufficiently large t
$$|z(t)|^2 \leq c \exp (2(\lambda_n + \beta + a) t)$$
and, consequently, for sufficiently small a
$$\pi(z) \leq \lambda_n + \beta + a, \qquad \pi(x) \leq \lambda_n + a < 0$$
This proves the asymptotic stability of the equilibrium of (26.4) and the first part of the theorem. Now, if the largest order number λ_n is positive, let $z = DY^T x$. As above, but with $\beta = 0$, we obtain
$$\pi(z) \leq \pi(x) \qquad (27.3)$$
The component z_n satisfies the differential equation
$$\dot{z}_n = \lambda_n z_n + \sum_{i=1}^{n} g_i y_{in} e^{\lambda_n t}$$
from which
$$z_n = c e^{\lambda_n t} + e^{\lambda_n t} \int_{t_0}^{t} \sum_{i=1}^{n} g_i y_{in} \, dt$$
follows. If the equilibrium were stable, the functions g_i would be bounded for sufficiently small initial values; the order number of the integral would be $-\lambda_n$ and hence the order number of the second term would be nonpositive, so that

by (22.2): $\qquad \pi(z_n) = \lambda_n > 0$

and by (27.3): $\qquad \pi(x) \geq \lambda_n > 0$

follows in contradiction to the assumption of stability. Consequently, the equilibrium of (26.4) is unstable.

A slight modification of the proof leads to the following:

Corollary (Chetaev [7]): Assume that the differential equation (26.5) of the first approximation is not regular. If all its order numbers are smaller than the always negative quantity (cf. (25.5))
$$-\kappa = -\pi(x_1) - \cdots - \pi(x_n) - \pi(1/d)$$
then the equilibrium of (26.4) is asymptotically stable; if at least one of the order numbers is greater than κ, the equilibrium of (26.4) is unstable. If the inequality
$$|g(x, t)| \leq a|x|^r \quad \text{with} \quad r > 1$$
is satisfied, then the condition
$$\lambda_n < -\kappa/(r-1)$$
is sufficient for asymptotic stability of the equilibrium of (26.4) (Massera [4]).

The criteria for the stability according to the first approximation formulated in the theorems of Secs. 26 and 27 are not equivalent: The regularity of a linear differential equation is neither necessary nor sufficient for intensive behavior of its motions, which can be seen from counter examples (Perron [4], Malkin [19]). A criterion has been formulated by Maizel [1] which comprises the theorems just mentioned; but it only states sufficient conditions. A theorem comprising the necessary and sufficient conditions for the general case of stability according to the first approximation is not yet known. Chetaev [11] gave an instructive survey of this group of problems.

28. TOTAL STABILITY

In Sec. 26 it was assumed that the differential equations to be compared, namely (26.1) and (26.2), both have the trivial solution; i.e., in particular, that $g(0, t) \equiv 0$. If (26.2) is considered as the equation of motion for a real, physical system upon which certain small perturbation forces act, described by $g(x, t)$, it must be realized that often such forces are not accurately known; in general, there are at most estimates for the perturbations. Thus, the assumption $g(0, t) \equiv 0$ is an idealization not justified by physical reality. If this assumption is dropped, the "stability" statement that small perturbation forces $g(x, t)$ produce only small deviations cannot be formulated by means of the stability concept defined in Sec. 2; an extension of this concept is needed.

The right hand side of the differential equation

$$\dot{x} = f(x, t) \qquad (28.1)$$

as well as the right hand side of the "perturbed" differential equation

$$\dot{x} = f(x, t) + g(x, t) \qquad (28.2)$$

shall be continuous and shall belong to the class E. Furthermore, let $f(0, t) \equiv 0$ for $t \geq t_0$.

DEFINITION 28.1: The equilibrium $x = 0$ of the differential equation (28.1) is called *totally stable*, if for every $\epsilon > 0$ two positive numbers $\delta_1(\epsilon)$ and $\delta_2(\epsilon)$ can be found such that for every solution $p(t, x_0, t_0)$ of (28.2), the inequality

$$|p(t, x_0, t_0)| < \epsilon \qquad (t > t_0)$$

holds, provided that

$$|x_0| < \delta_1$$

and

$$|g(x, t)| < \delta_2 \qquad (28.3)$$

in the domain $\Re_{\epsilon, t_0}$.

In the Russian literature this type of stability is called *stability under constantly acting perturbations*. The concept was introduced by Dubošin [4] with proper modification of Definition 28.1 for an arbitrary motion of

(28.1). Vorovich [1] recently extended the definition to the case where the perturbation forces are statistically given random functions.

Theorem 28.1: *If the equilibrium of (28.1) is uniformly asymptotically stable, then it is also totally stable.*

The proof of this theorem was given in different, but equivalent, forms by Goršin [1] and Malkin [9, 19]. Goršin's considerations seem to be independent of those of Malkin. Antosiewicz [1] gave another proof. Malkin's proof is based on the following:

Theorem 28.2 (Malkin [9]): If a positive definite Liapunov function $v(x, t)$ exists whose partial derivatives are bounded in a domain and whose total derivative for (28.1) is negative definite, then the equilibrium is totally stable.

Proof: According to Theorem 1.2, v is decrescent. Thus there exist three functions $\varphi(r)$, $\psi(r)$ and $\chi(r)$ of the class K such that

$$\varphi(|x|) \leq v(x, t) \leq \psi(|x|), \quad \dot{v}_{(28.1)} \leq -\chi(|x|)$$

Let $\epsilon > 0$ be given and let $0 < \beta < \varphi(\epsilon)$. Then there exists a number $\gamma = \gamma(\beta) > 0$ such that from $v(\tilde{x}, t) = \beta$ the inequality

$$\gamma < |\tilde{x}| < \epsilon$$

follows and, in addition,

$$\dot{v}_{(28.1)}(\tilde{x}, t) < -\chi(\gamma) \tag{28.4}$$

The functions $\dot{v}_{(28.1)}$ and $\dot{v}_{(28.2)}$ differ by the occurence of the term

$$\sum_{i=1}^{n} g_i \frac{\partial v}{\partial x_i}$$

which (because of the additional assumption) can be made arbitrarily small by suitable choice of δ_2 in (28.3). Thus with sufficiently small δ_2, we also have $\dot{v}_{(28.2)} < 0$ in the domain $|x| < \delta_2$. Now, choose $\delta_1 = \delta_1(\epsilon)$ such that

$$\delta_1 < \epsilon \quad \text{and} \quad v(t_0) = v(x_0, t_0) < \beta \quad \text{if} \quad |x_0| < \delta_1$$

Then

$$|p(t, x_0, t_0)| < \epsilon \quad (t > t_0)$$

If this inequality were not valid for $t_1 > t_0$, we would have $v(t_1) > \beta$ since $\beta < \varphi(\epsilon)$. But $v(t_0) < \beta$ and, since $\dot{v}_{(28.2)} < 0$, the function $v(t)$ decreases monotonically.

Theorems 18.3 and 28.2 together yield Theorem 28.1.

If the differential equation (28.2) is autonomous and if its equilibrium is asymptotically stable, then the domain of attraction of the equilibrium is, in a certain sense, insensitive to perturbations. A function $g(x)$, independent of t, can be found such that the equilibrium of (28.2) has the same domain of attraction as the equilibrium of (28.1), provided that

$$|g(x, t)| < g(x)$$

(Evidently, it must be required that $g(0, t) \equiv 0$.) Zubov [6] proved this theorem by means of the methods of Sec. 21. Massera [2] considered the periodic solution of a differential equation with periodic coefficients under the assumption that the periods of the solution and of the equation are incommensurable. Massera stated conditions for the total stability of this solution.

Krasovskii [6] introduced the concept of *total stability in the whole* for autonomous differential equations. In addition to the conditions of Definition 28.1 it is required that for every $\epsilon > 0$ there exists a number $\eta(\epsilon) > 0$ such that for every solution $x(t)$ of (28.2) lim sup $|x| < \epsilon$ $(t \to \infty)$, provided that in the domain $\Re_{\epsilon, t_0}$ an inequality of the form

$$|g| < \eta \kappa(|x|)$$

holds. In this inequality, $\kappa(r)$ is a given positive function which characterizes the growth of the perturbation function in the phase space. A general condition for total stability in the whole is not yet known. Krasovskii [6] restricted his investigations to the derivation of sufficient conditions for the differential equations treated in Sec. 14, under (b) and (c). He also considered the system

$$\dot{x} = f_1(x, y), \quad \dot{y} = f_2(s), \quad s = ax - by$$

which has been investigated before by Eršov [1, 2]. This system admits a control engineering interpretation.

The Insensitivity Theorem 28.1 remains valid if, with regard to the perturbation functions, only "smallness in the average" is assumed in place of (28.3). In this case, the existence of a continuous function $\varphi(t)$ must be required which satisfies in the domain $\Re_{\epsilon, t_0}$ the inequalities

$$\int_t^{t+\tau} \varphi(u) \, du < \delta_2, \qquad |g(x, t)| < \varphi(t) \tag{28.5}$$

Here, τ is an arbitrary positive number. The bounds δ_1 and δ_2, defined in Definition 28.1, may depend on τ (Germaidze and Krasovskii [1]). Under very general assumptions on the right hand side of (28.1), Vrkoč [2] closely investigated the case in which the perturbation function $g(x, t)$ in (28.3) satisfies certain integral conditions (integral stability).

The criteria mentioned in Sec. 4, Remark 5, and Sec. 17, Remark 1, which are based on a differential equation for v, can also be expressed for the previously formulated insensitivity statement: The equilibrium of (28.1) is totally stable, or integrally stable (in the sense of Vrkoč, or Germaidze-Krasovskii) if the corresponding is true for the equilibrium

$y = 0$. In addition, a Lipschitz condition for v must be satisfied (Corduneanu [3]).

Theorem 28.1 cannot be inverted (Massera [4], Erratum). However, the following is valid:

THEOREM 28.3 (Massera [4]): If the equilibrium of the linear differential equation

$$\dot{x} = A(t)\, x$$

is totally stable, then it is uniformly asymptotically stable (and, by Theorem 24.3, also exponentially stable).

Proof: Because of the total stability, the equilibrium of the linear perturbed auxiliary equation

$$\dot{y} = (A(t) + \eta U)\, y$$

is stable for sufficiently small $\eta > 0$. However,

$$x(t) = e^{-\eta t} y(t)$$

from which the uniformity of the asymptotic stability follows.

7 THE CRITICAL CASES

29. GENERAL REMARKS ON THE CRITICAL CASES

A *critical case* exists if the equilibrium of the linearized differential equation of a given complete equation is stable, but not exponentially stable (cf. Sec. 8 and Sec. 26). In this case, the stability behavior of the complete differential equation is no longer determined solely by the linear terms of the power series expansions of the right hand sides; it depends rather on the terms of higher order. Consequently, the critical cases are not actually special cases; they are characterized by the fact that a particular special case is not present. From this it follows that the theory of such cases does not involve general criteria such as, for example, that of stability according to the first approximation. Although criteria could, in principle, be established for a theory of "stability according to the second approximation," etc., the numerous necessary discriminations did limit the investigations to the consideration of some special differential equations. The few general theorems do not actually represent criteria; they contain assumptions under which stability statements can be made by means of a finite process, for example, by constructing a differential equation with the same stability behavior as the original equation, but of simpler structure. Since criticial cases occur quite frequently in concrete problems of physics and mechanics, such reduction procedures are of considerable importance for applications, provided they are reasonably amenable.

In the following sections, the simplest critical cases of autonomous differential equations will be treated. Some of these cases have already been solved by Liapunov [1]. Several approaches are available in general to solve the stability problem. In the sequel, always that procedure in

112   THE CRITICAL CASES   CHAP. 7

which the direct method is used will be sketched. A presentation of the theory in form of a textbook was given by Malkin [19] (cf. also Dubošin [7], Zubov [6]). In order to arrive at a classification of the critical cases for autonomous differential equations, the following can be done. Let

$$\dot{x} = Ax + g(x) \qquad (g \in C_\omega, \text{ cf. Sec. 1(e)}) \tag{29.1}$$

be the given differential equation, written in the form (26.4). The real parts of the eigenvalues of A shall be nonpositive. Those eigenvalues which have vanishing real parts are called *critical;* their number determines the degree of difficulty of the investigation. One eigenvalue is critical in the simplest case; i.e., the characteristic equation of A has one simple vanishing root. The following can occur in the case of two critical eigenvalues: (a) two purely imaginary roots which are conjugate complex, (b) a vanishing double root to which an elementary divisor of second order corresponds, and (c) a vanishing double root with two elementary divisors of first order. In the case of three critical eigenvalues, there are four subcases, etc.

In order to solve the stability problem, the differential equation is at first subjected to a linear transformation such that the variables split into two groups, corresponding to *critical* and *noncritical* eigenvalues. Then certain nonlinear transformations are applied which, in a sense, eliminate the effect of the noncritical variables without changing the stability of the equilibrium. Thus, a differential equation is obtained which is equivalent to the original equation with respect to stability but which contains only as many variables as there are critical eigenvalues. Finally a suitable Liapunov function is constructed for this *reduced* differential equation which enables one to decide upon the stability of the equilibrium.

30. THE TWO SIMPLEST CRITICAL CASES

(a) The matrix A of (29.1) shall have the single critical eigenvalue 0. By the linear transformation, mentioned at the end of Sec. 29 and by a suitable renotation of the variables, the differential equation changes into the form

$$\dot{y}_0 = g_0(y_0, y_1, \ldots, y_n) \tag{30.1}$$

$$\dot{y}_i = \sum_{j=1}^{n} p_{ij} y_j + p_i y_0 + g_i(y_0, y_1, \ldots, y_n) \qquad (i = 1, \ldots, n) \tag{30.2}$$

The power series expansions of g_0, g_i begin with terms of at least second order. The equations

$$\sum_{j=1}^{n} p_{ij} y_j + p_i y_0 + g_i(y_0, y_1, \ldots, y_n) = 0 \qquad (i = 1, \ldots, n)$$

can be solved with respect to the variables $y_1, \ldots, y_n$, since the functional determinant does not vanish at the point $(0, \ldots, 0)$ as a consequence of det $(p_{ij}) \neq 0$. Its solution shall be denoted by

$$u_1(y_0), \ldots, u_n(y_0)$$

The substitution

$$y_0 = z_0, \qquad y_i = z_i + u_i(y_0) \qquad (i = 1, \ldots, n)$$

transforms (30.1) and (30.2) into

$$\dot{z}_0 = h_0(z_0, z_1, \ldots, z_n) \tag{30.3}$$

$$\dot{z}_i = \sum_{j=1}^{n} p_{ij} z_j + h_i(z_0, z_1, \ldots, z_n) \qquad (i = 1, \ldots, n) \tag{30.4}$$

In these equations, the function h_i contains the term

$$\dot{u}_i = \frac{du_i}{dz_0} \dot{z}_0 = \frac{du_i}{dz_0} h_0 \tag{30.5}$$

The differential equations (30.4) are characterized by the fact that the critical variable z_0 no longer occurs linearly. Furthermore, the term (30.5) causes the order of z_0 in (30.3) to be not larger than the order of z_0 in the Eqs. (30.4). However, the assumption must be made that the function $h_0(z_0, z_1, \ldots, z_n)$ does not vanish identically.

Let the expansion of the expression $h_0(z_0, 0, \ldots, 0)$ in powers of z_0 begin with the term

$$g z_0^m \qquad (m \geq 2) \tag{30.6}$$

Then the following theorem is valid:

THEOREM 30.1 (Liapunov [1]): Let the assumptions just formulated for the differential equations (30.3) and (30.4) be satisfied. If m is an even number, the equilibrium $z = 0$ is unstable; consequently, the equilibrium of (30.1; 2), or of (29.1), respectively, is unstable. If m is odd and $g > 0$, the result is the same. If m is odd and $g < 0$, the equilibrium is asymptotically stable.

Proof: The proof can be given by means of the direct method as follows. If there is only a single equation $(n = 0)$, the Liapunov function

$$v = g z_0^2 \qquad \text{(for odd } m\text{)}$$
or
$$v = g z_0, \qquad \text{respectively (for even } m\text{)} \tag{30.7}$$

is chosen. Its derivative

$$\dot{v} = 2g^2 z_0^{m+1} + \cdots, \quad \text{respectively} \quad \dot{v} = g^2 z_0^m + \cdots$$

is always positive definite, and the statement of the theorem follows from Theorem 4.1, or 5.2, respectively. If $n \geq 1$, a Liapunov function v_1 is obtained by adding to (30.7) additional terms, namely:

1. the quadratic form $z^T B z$, taken with the negative sign, which is formed in accordance with the procedure of Sec. 8 for the linear system

$$\dot{z}_i = p_{i1} z_1 + \cdots + p_{in} z_n \qquad (i = 1, \ldots, n)$$

2. a linear combination of $z_0^2, z_0^3, \ldots, z_0^m$, whose coefficients are certain linear forms of the noncritical variables $z_1, \ldots, z_n$. The coefficients are to be chosen such that $\dot{v}_1$ is definite. It can be shown that these coefficients are uniquely determined by these requirements. The Liapunov function so constructed has the same properties as the function v defined by (30.7).

The exceptional case, excluded above, is solved by the following:

THEOREM 30.2 (Liapunov [1]): If the function h_0 of (30.3) vanishes identically, then the equilibrium is weakly stable. In this *singular* case, the equilibrium belongs to a family of stationary motions $z_0 = \text{const}$, $z_i = 0$, or $y_0 = \text{const}$, $y_i = u_i(y_0)$, respectively, $(i = 1, \ldots, n)$.

This theorem can be generalized.

THEOREM 30.3 (Liapunov [1], cf. also Malkin [6]): Let two differential equations be given for the k-rowed vector y and the n-rowed vector x:

$$\dot{y} = g(x, y, t), \qquad \dot{x} = Px + h(x, y, t)$$

The power series expansions of the components of g and h, with respect to the variables x_i, y_i, shall begin with terms of at least second order. Furthermore, let

$$g(0, y, t) \equiv 0, \qquad h(0, y, t) \equiv 0$$

and let the constant matrix P have only eigenvalues with negative real parts. Then the equilibrium is weakly stable, and every solution sufficiently close to the equilibrium tends, with increasing t, towards a stationary solution of the form

$$y = c (= \text{const}), \qquad x = 0$$

Theorem 30.1 permits a solution of the stability problem for the differential equation $\dot{x} = f(x)$ in the critical case. But as a consequence of the complicated procedure of the proof, it does not make it very clear how the stability behavior of the equilibrium is determined by the structure of the differential equation. This question is answered by:

THEOREM 30.4 (Krasovskii [11]): If the Jacobian matrix $(\partial f_i / \partial x_k)$ has the simple eigenvalue 0 at the point $x = 0$, and if in a neighborhood of this point it has only eigenvalues with negative real parts, then the equilibrium is asymptotically stable. However, if at every point of a neighborhood of $x = 0$, this point excluded, there is an eigenvalue with positive real part, then the equilibrium is unstable.

The proof, which was given by Krasovskii [5] for the special case $n = 2$ and later for the general case, is also based on a nonlinear transformation which results in equations of a form analogous to (30.3) and (30.4). For these equations a suitable Liapunov function is constructed. Difficulties arise in the discussion of the sign of that term which corresponds to (30.6).

(b) Let the matrix A of (29.1) have two purely imaginary eigenvalues $\pm i\lambda$. The differential equation can then be put into the form

$$\dot{\mathsf{y}} = \mathsf{L}\mathsf{y} + \mathsf{g}(\mathsf{y}, \mathsf{x}), \qquad \dot{\mathsf{x}} = \mathsf{P}\mathsf{x} + \mathsf{Q}\mathsf{y} + \mathsf{h}(\mathsf{y}, \mathsf{x}) \qquad (30.8)$$

Here, y is a 2-rowed vector, x an n-rowed vector, and

$$\mathsf{L} = \begin{pmatrix} 0 & -\lambda \\ +\lambda & 0 \end{pmatrix}$$

First, let $n = 0$; then only the critical variables y_1, y_2 occur. The first equation of (30.8) is written in the form

$$\dot{\mathsf{y}} = \mathsf{L}\mathsf{y} + \mathsf{g}^{(2)}(\mathsf{y}) + \mathsf{g}^{(3)}(\mathsf{y}) + \cdots \qquad (30.9)$$

All terms in which the variables are of the kth order are contained in the function $\mathsf{g}^{(k)}(\mathsf{y})$. The Liapunov function is written as a series

$$v = y_1^2 + y_2^2 + f_3(\mathsf{y}) + f_4(\mathsf{y}) + \cdots \qquad (30.10)$$

which correspondingly progresses by forms of second, third, ... order. In particular, these forms are so determined that the total derivative $\dot{v}$ for (30.9) begins with a term of the form

$$\gamma_{2m}(y_1^2 + y_2^2)^m \qquad (m \geq 2)$$

Equations for the forms f_i can be obtained; the first two of these equations are

$$\lambda \left(y_1 \frac{\partial f_3}{\partial y_2} - y_2 \frac{\partial f_3}{\partial y_1} \right) = -2y_1 g_1^{(2)} - 2y_2 g_2^{(2)}$$

$$\lambda \left(y_1 \frac{\partial f_4}{\partial y_2} - y_2 \frac{\partial f_4}{\partial y_1} \right) = -f_4^*(\mathsf{y}) + \gamma_4(y_1^2 + y_2^2)^2$$

where

$$f_4^*(\mathsf{y}) = \frac{\partial f_3}{\partial y_1} g_1^{(2)} + \frac{\partial f_3}{\partial y_2} g_2^{(2)} + 2y_1 g_1^{(3)} + 2y_2 g_2^{(3)}$$

$$\gamma_4 = \frac{1}{2\pi} \int_0^{2\pi} f_4^* (\cos \varphi, \sin \varphi) \, d\varphi$$

These partial differential equations determine uniquely the forms f_3 and f_4 (Liapunov [1], Malkin [7, 19]). If $\gamma_4 \neq 0$, we put

$$v = y_1^2 + y_2^2 + f_3 + f_4 \qquad (30.11)$$

Then

$$\dot{v} = \gamma_4(y_1^2 + y_2^2)^2 + \text{terms of higher order.}$$

It can be seen that the equilibrium is unstable in the case where $\gamma_4 > 0$, and asymptotically stable in the case where $\gamma_4 < 0$.

If γ_4 vanishes, we have to proceed by computing f_5, f_6, or even higher forms. But if all γ_{2m} vanish, then $\dot v$ is identically zero, the equilibrium is weakly stable, and the curves $v = $ const are integral curves. In this case, the origin is a center.

If the differential equation (29.1) contains also noncritical variables, we try to eliminate their effect in the same manner as in (a). Here again, a nonlinear transformation of the variables

$$z = x - v(y_1, y_2)$$

is applied. The components of v are determined by a system of partial differential equations:

$$\frac{\partial v_i}{\partial y_1}(-\lambda y_2 + g_1(y, v)) + \frac{\partial v_i}{\partial y_2}(\lambda y_1 + g_2(y, v))$$
$$= p_{i1}v_1 + \cdots + p_{in}v_n + q_{i1}y_1 + q_{i2}y_2 + h_i(y, v) \qquad (i = 1, \ldots, n)$$

The terms which are linear in the critical variables are eliminated by the transformation, and the orders in which the critical variables occur in the terms of h are not smaller than the corresponding orders in g. A Liapunov function is obtained for the transformed differential equations by adding suitable additional terms to (30.10), or (30.11), respectively, if a nonvanishing number γ_{2m} should occur. The sign of this number is decisive for the stability behavior of the equilibrium.

On the other hand, if all $\gamma_{2m} = 0$, then the equilibrium is weakly stable, and the original differential equation has periodic solutions. This special case is obtained, for instance, if the differential equation (30.1) has complex coefficients, and if the assumption is made that the matrix has a single, purely imaginary eigenvalue. The equilibrium of the equivalent real system is weakly stable and has periodic solutions (Vejvoda [1]).

(c) As already mentioned, the solution of the stability problem in a critical case can be approached in different ways. The approach sketched in (a) and (b) is characterized by its methodical clarity, but it is not very well suited for practical application since the single steps, in particular the computation of the components of v, are in most cases very cumbersome; Malkin [11, 12] developed a procedure which considerably simplifies the computations by applying several tricks. It is evident that an answer to the stability question for many particular cases can be obtained quicker by direct construction of a Liapunov function. Berezkin [1] studied in this manner some differential equations of the type

$$\dot x = y(1 + f(x)), \quad \dot y = \varphi(x) + y\varphi_1(x) + \cdots \quad (f(0) = 0)$$

For instance, if $\varphi(x) = 0$, $\varphi_1(x) \neq 0$ for $0 < x < c$ it can be seen by means of the Liapunov function $v = xy$ that the equilibrium is unstable.

A geometrical interpretation of the procedure described in (a) and (b) was given by Moisseev [3].

31. MALKIN'S COMPARISON THEOREMS

The decisive step in the procedures of the preceding section is the introduction of the nonlinear transformation of variables. In a well determined manner, the transformation increases the minimum order of the critical variables occurring in the individual differential equations. Thereby it becomes possible to consider a *reduced* system instead of the original system of differential equations, which contains only as many (scalar) differential equations as there are critical variables. A Liapunov function v, formed for the reduced system, can be transformed into a Liapunov function v_1 for the complete system by adding certain additional terms. The functions v and v_1 on the one hand and their derivatives $\dot{v}$ and $\dot{v}_1$ for the corresponding differential equations on the other have the same definiteness properties. Hence, it can be concluded that the complete system and the reduced system are equivalent with respect to the stability behavior of their equilibria.

Malkin [8, 19] recognized that the considerations of Sec. 30 are special cases of a much more general situation. The generalization refers to the following three points:

1. The number of critical variables is not confined to one or two;
2. the differential equations are not necessarily autonomous;
3. stability according to the first approximation can be replaced by *stability according to the mth approximation*. Here, mth approximation means that the stability behavior of the equilibrium depends only on those terms in the series expansions of the right hand sides whose orders are not greater than m and that the stability behavior is in no way effected by terms of higher order.

Let the differential equation

$$\dot{z} = f^{(r)}(z, t) + f^{(r+1)}(z, t) + \cdots + f^{(m)}(z, t) + \mathfrak{g}(z, t) \qquad (31.1)$$

be given in $\mathfrak{K}_{h,t_0}$. Let the components of $f^{(k)}(z, t)$ be forms of the order k in the variables $z_1, \ldots, z_n$ with continuous and bounded coefficients. The term $\mathfrak{g}(z, t)$ comprises all those terms whose order exceeds the number m so that

$$|\mathfrak{g}(z, t)| < a|z|^m \qquad (a = \text{const}) \qquad (31.2)$$

DEFINITION 31.1: The equilibrium of (31.1) is said to be *stable according to the mth approximation* if for every $\epsilon > 0$ a number $\delta > 0$, depending only on ϵ and a, but not on $\mathfrak{g}(z, t)$, can be determined such that from $|z(t_0)| < \delta$ the inequality

$$|z(t)| < \epsilon \qquad (t > t_0)$$

follows, independent of the particular form of the function $\mathfrak{g}(z, t)$, provided only the inequality (31.2) is satisfied. If even *in addition*

$$\lim_{t \to \infty} |z(t)| = 0$$

the equilibrium is called *asymptotically stable according to the mth approximation*. If under the same assumption on $g(z, t)$ there is a number $\epsilon > 0$ (depending on a) and a number $\tau = \tau(\epsilon)$ such that for arbitrarily small δ and for arbitrary choice of $g(z, t)$ there is at least one motion $\bar{z}(t)$ with $|\bar{z}(t_0)| < \delta$ which satisfies the relation $|\bar{z}(t + \tau)| \geq \epsilon$, then the equilibrium is called *unstable according to the mth approximation*.

In order to formulate Malkin's theorems, the original differential equation is written in a form similar to that of Sec. 30:

$$\begin{aligned} \dot{y} &= R(t) x + g(y, x, t) \\ \dot{x} &= P(t) x + h(y, x, t) \quad (t \geq t_0 \geq 0, |x| \leq h, |y| \leq h) \end{aligned} \qquad (31.3)$$

Here, y is a k-rowed vector, x an n-rowed vector. The matrices P and R shall be continuous and bounded. The series expansions of the components of g and h with respect to y and x shall begin in the domain $\mathfrak{K}_{h,t_0}$ with terms of at least second order. Let the equilibrium of the linear auxiliary system

$$\dot{x} = P(t) x \qquad (31.4)$$

be exponentially stable. Besides (31.3), the *reduced* equation

$$\dot{y} = h(y, 0, t) \qquad (31.5)$$

is considered. Under these assumptions the following theorem is valid:

THEOREM 31.1 (Malkin [8, 19]): Let the equilibrium of (31.5) be asymptotically stable, weakly stable, or unstable, according to the mth approximation. If the series expansions of $h(y, 0, t)$ begin with terms of order $\alpha \geq m + 1$, then the equilibrium of the complete differential equation (31.3) has the same stability behavior as the equilibrium of the reduced equation.

If the first differential equation of (31.3) contains an additional term $Q(t) y$, where $Q(t)$ is bounded, which then also appears in (31.5), then Theorem 31.1 is valid only under the additional assumption that the system of the partial differential equations

$$\frac{\partial u_i}{\partial t} + \sum_{s=1}^{n} \left[\sum_{k=1}^{n} p_{sk} x_k + h_s(t, u, x) \right] \frac{\partial u_i}{\partial x_s}$$
$$= \sum_{j=1}^{k} q_{ij} u_j + g_i(u, x, t) \qquad (i = 1, \ldots, k)$$

admits an analytic solution u, which is bounded with respect to t. (The form in which Theorem 31.1 is expressed here is quoted from the German edition of Malkin [19].)

Instead of (31.3), we consider the new system

$$\dot{y} = Q(t)\,y + g(y, x, t)$$
$$\dot{x} = P(t)\,x + R(t)\,y + h(y, x, t) \qquad (31.6)$$

For this new system, the assumptions of Theorem 31.1 are not satisfied. In this case, we try to satisfy these assumptions by means of a nonlinear transformation

$$x = z + u(y) \qquad (31.7)$$

Formal power series

$$u_i = \Sigma\, a_i^{(m_1,\ldots,m_k)}(t)\, y_1^{m_1} y_2^{m_2} \cdots y_k^{m_k} \qquad (m_1 + \cdots + m_k \geq 1) \qquad (31.8)$$

with bounded coefficients $a_i(t)$ are chosen for the components u_i of u such that the partial differential equations

$$\frac{\partial u_s}{\partial t} + \sum_{i=1}^{n} \frac{\partial u_s}{\partial y_i}(q_{i1}y_1 + \cdots + q_{ik}y_k + g_i(y, u, t))$$
$$= p_{s1}u_1 + \cdots + p_{sn}u_n$$
$$+ r_{s1}y_1 + \cdots + r_{sk}y_k + h_s(y, u, t) \qquad (s = 1, \ldots, n) \qquad (31.9)$$

are formally satisfied. (The left hand side represents the expression du_s/dt for (31.6) with u in place of x.) The word "formal" shall indicate that the convergence of the series under consideration is unimportant. If the quantity x in the first equation of (31.6) is now replaced by u(y), then the reduced equation

$$\dot{y} = Q(t)\,y + g(y, u(y), t) \qquad (31.10)$$

is obtained, and the following theorem holds:

THEOREM 31.2 (Malkin [8, 19]): If the equilibrium of the reduced differential equation (31.9) is asymptotically stable, weakly stable, or unstable, according to the mth approximation (m arbitrary, but finite), then the equilibrium of the complete equation has corresponding behavior. Here it is assumed that the differential equations (31.9) can be solved by power series with bounded coefficients.

A Liapunov function for the linear auxiliary system (31.4) is used to prove Theorem 31.1. In order to apply Theorem 31.2, the partial differential equations (31.9) must be solved by power series (31.8) with bounded coefficients. The conditions for this being possible are not yet completely known; however, the solvability has been proved for some important special cases. This is expressed in:

THEOREM 31.3 (Malkin [7, 8, 19]): The differential equations (31.9) possess formal solutions of the form (31.8) with bounded coefficients (at least) under the following conditions: (a) the matrix $Q(t)$ vanishes identically; (b) the matrices P, Q, R are constant. In the case (b) there exist even

120   THE CRITICAL CASES   CHAP. 7

infinitely many solutions. If here g and h are periodic in t, then there is exactly one formal solution with periodic coefficients; in this case, the reduced differential equation has also periodic coefficients. If g and h are independent of t, then there exists exactly one solution of the differential equations which does not depend on t; the corresponding reduced differential equation is autonomous.

32. SPECIAL INVESTIGATIONS OF CRITICAL CASES

(a) The procedure treated in Sec. 31 is predominantly of theoretical value, since the computations leading to the reduced differential equation are generally rather tedious, even disregarding the stability investigation which has to be carried out for the reduced equation. However, as already mentioned at the end of Sec. 30, in some special cases the procedure can be abbreviated and the stability question can be answered completely. In particular, Malkin [2, 6, 19] solved the case where the matrix A of (29.1) has a double eigenvalue 0 with two associated elementary divisors of the first order. He also showed (Malkin [13, 19]) that two other critical cases can be reduced to this case, namely:

(α) the matrix has two distinct pairs of purely imaginary eigenvalues;
(β) the matrix has a simple eigenvalue zero, and one pair of purely imaginary eigenvalues.

Beyond this, Zubov [6] studied the case of a multiple eigenvalue 0, as well as the case of more than two pairs of imaginary eigenvalues.

(b) For the reduced differential equation (with two scalar components) belonging to the first special case, the stability behavior can be described by a *Rule* (Malkin [2, 6, 19]; cf. also Kamenkov [1]). In analogy to (30.9), we write the differential equation in the form

$$\dot{y} = g^{(m)}(y_1, y_2) + g^{(m+1)}(y_1, y_2) + \cdots \qquad (m \geq 2) \qquad (32.1)$$

The functions $g_i^{(k)}(y_1, y_2)$ ($i = 1, 2$) are forms of the order k. Furthermore, let

$$f_1(y) = y_1 g_1^{(m)} + y_2 g_2^{(m)}, \qquad f_2(y) = y_1 g_2^{(m)} - y_2 g_1^{(m)}$$

We introduce the quantity

$$\alpha = \frac{1}{2\pi} \int_0^{2\pi} \frac{f_1(\cos \varphi, \sin \varphi)}{f_2(\cos \varphi, \sin \varphi)} d\varphi \qquad (32.2)$$

and we consider the straight lines of the (y_1, y_2)-plane defined by the equation

$$f_2(y_1, y_2) = 0 \qquad (32.3)$$

Then the following *Rule* holds:

1. The equilibrium of (32.1) is asymptotically stable if f_2 is not definite

and if f_1 takes on only negative values along all straight lines (32.3) except at the origin.

2. The equilibrium is unstable if f_2 is not definite and if f_1 takes on positive values along at least one of the straight lines (32.3).

3. The equilibrium is asymptotically stable if f_2 is definite, if $\alpha \neq 0$, and if $\alpha f_2 < 0$; the equilibrium is unstable if f_2 is definite, if $\alpha \neq 0$, but if $\alpha f_2 > 0$.

4. Higher terms of the expansion (32.1) must be taken into account in order to solve the stability problem if f_2 is definite and if $\alpha = 0$, or if f_2 is not definite and if f_1 vanishes along some of the straight lines (32.3) without taking on positive values.

If differential equation (29.1) has complex coefficients and if A has a critical eigenvalue zero, then, by the Comparison Theorems, the stability problem can be reduced to the investigation of the scalar differential equation

$$\dot{z} = cz^k + \text{terms of higher order} \qquad (c \text{ complex}, k \geq 2) \qquad (32.4)$$

From case 2 of the Rule just mentioned, it follows that the equilibrium of (32.4) is always unstable. This can be proved directly in a simpler manner (Vejvoda [1]).

(c) If the matrix A in (29.1) has a double eigenvalue 0 and if the corresponding elementary divisor has the order 2, the investigation becomes considerably more complicated. This case was solved by Liapunov [2] for the reduced differential equation and by Kamenkov [2] for the complete equation. If the reduced equation is written in the form

$$\dot{y}_1 = y_2 + g(y_1, y_2)$$
$$\dot{y}_2 = ay_1^\alpha + a'y_1^{\alpha+1} + \cdots + y_2(by_1^\beta + b'y_1^{\beta+1} + \cdots) + y_2^2 h(y_1, y_2)$$

(g is at least of second order, α and β are positive integers), then the stability question can be answered by the following *Criteria* of Liapunov [2]:

The equilibrium is *unstable*, if one of the following conditions is satisfied:

1. α even;
2. α odd, $a > 0$;
3. α odd, $a < 0$, β even, $\alpha \geq \beta + 1$, $b > 0$;
4. α odd, $a < 0$, β odd, $\alpha \geq 2\beta + 2$;
5. α odd, $a < 0$, β odd, $\alpha = 2\beta + 1$, $b^2 + 4(\beta + 1)a \geq 0$;
6. the equation for $\dot{y}_2$ contains only the term $h(y_1, y_2)$.

The equilibrium is *asymptotically stable* if

7. α odd, $a < 0$, β even, $\alpha \geq \beta + 1$, $b < 0$;
8. β even, $b < 0$, $ay_1^\alpha + a'y_2^{\alpha+1} + \cdots \equiv 0$.

Higher terms of the expansion must be considered in the following cases:

9. α odd, $a < 0$, β even, $\alpha < \beta$;
10. α odd, $a < 0$, β odd, $\alpha < 2\beta$;
11. α odd, β odd, $\alpha = 2\beta + 1$, $b^2 + 4(\beta + 1)a < 0$.

Berezkin [2] considered the following specific system

$$\dot{x} = y, \quad \dot{y} = (b_1 - b_2 y)z^2, \quad \dot{z} = -ax - by - cz$$

The equilibrium is unstable if all constants are positive, or if $a = 0$, $b_1 > 0$. The equilibrium is weakly stable if $a > 0$, $b_1 = 0$.

(d) We now consider a differential equation with periodic coefficients. Here, a critical case occurs if there are no roots of the characteristic equation of the linearized differential equation (cf. Sec. 24) of absolute value greater than 1, but if there is at least one root whose absolute value is equal to 1. With regard to the theorems of Sec. 31, it suffices to consider only the reduced differential equation

$$\dot{y} = Q(t)y + g(y, t) \qquad (32.5)$$

The vector y contains as many variables as there are critical roots. By means of the Reducibility Theorem of Liapunov (Sec. 24), Eq. (32.5) can be transformed into an equation whose linear part has a constant coefficient matrix; it may readily be assumed that this matrix is the Jordan normal form J. Thus, (32.5) can be written in the form

$$\dot{z} = Jz + f^{(2)}(z, t) + f^{(3)}(z, t) + \cdots + f^{(m)}(z, t) + g^*(z, t)$$

z is the vector of the new variables. The components of $f^{(r)}$ are forms of the order r in $z_1, \ldots, z_k$, and $|g^*| < a|z|^m$ with constant a. It may furthermore be assumed that the functions $f^{(2)}, \ldots, f^{(m)}$ have constant coefficients; this can always be achieved by means of a nonlinear transformation and by solving certain partial differential equations similarly as in Sec. 31. By this method, the stability problem for the periodic differential equation (32.5) is reduced to the corresponding problem for an autonomous differential equation.

Liapunov [1] solved the two simplest critical cases for periodic differential equations by reduction to autonomous equations. In these cases, the characteristic equations either have a simple root 1, or two conjugate complex roots of absolute value 1. No other cases have been treated in the literature since then. Kalinin [1, 2] somewhat simplified Liapunov's procedure by introducing polar coordinates and by replacing the time t by the polar angle. Aminov [2] investigated a special case of the differential equation (32.5) where the characteristic equation has the simple root 1, and where the differential equation has first integrals. He constructed a Liapunov function by means of these integrals (as in Sec. 7) which satisfies the assumptions of the theorem on weak stability.

Kalinin [3] investigated another special case, namely the equations

$$\ddot{x} + a\dot{x} + bx = -cy, \quad \dot{y} = ps + q(t)\,s^3, \quad s = x + \alpha x - hy$$

i.e., a system of the type (14.3) with periodic $q(t)$. The investigation leads to a critical case of a system with periodic coefficients.

If the differential equation of the perturbed motion for a periodic solution of an autonomous differential equation is determined according to (2.6), then a periodic differential equation is obtained whose linearized equation always has a periodic solution because of the autonomy of the original equation. The characteristic equation has at least one root 1, i.e., the system is critical. The stability problem for this case which is of interest with respect to practical applications is solved by:

THEOREM 32.1 (Andronov and Witt [1]): If the characteristic equation of the differential equation of the perturbed motion for the periodic solution of an autonomous equation has $n - 1$ roots whose absolute values are smaller than 1, then this periodic solution is (weakly) stable.

33. THE BOUNDARY OF THE STABILITY DOMAIN IN THE PARAMETER SPACE

Let the differential equation

$$\dot{x} = f(x, a, t) \qquad (f \in E) \tag{33.1}$$

be given where the function f depends continuously on parameters which are the components of the vector a.

Let the differential equation

$$\dot{x} = A(a, t)\, x \tag{33.2}$$

of the first approximation of (33.1) be either autonomous or periodic. Those values of the parameters for which (33.2) has an exponentially stable equilibrium form the *stability domain in the parameter space*. Estimates for this domain were given in Secs. 12 ff. The stability domain is bounded by a hyper-surface which is called the *stability boundary* and which has to be distinguished from the stability boundary in the space of the initial points. The stability boundary is characterized by the fact that the corresponding differential equations (33.1) have an equation of the first approximation (33.2) which has critical behavior. Thus, the stability boundary corresponds to those parameter values for which (33.2) has critical behavior. Analytically, the stability boundary can be characterized by the fact that the characteristic equation of (33.2) has roots with vanishing real parts, or of absolute values 1, respectively, but no roots with positive real parts, or absolute values greater than 1, respectively. The corresponding conditions can be written in closed form by means of the Hurwitz determinants (cf. Malkin [19]), and hence the equation of the stability boundary can be expressed in closed form.

For practical applications it is important to know how a physical system behaves if the parameter a leaves the stability domain. Let a lie on the boundary and let a + c be a neighboring point outside the boundary. The corresponding differential equation (33.1) has the form

$$\dot{x} = f(x, a, t) + g(x, a, c, t) \tag{33.3}$$

The additional term $g(x, a, c, t)$ takes on the value 0 for $c = 0$ and, because of the continuity with respect to a, it is certainly of the character of the additional term in (28.1). Thus, Theorem 28.1 can be applied which leads to:

THEOREM 33.1: If the equilibrium of (33.1) is asymptotically stable for a located on the stability boundary (i.e., if the equilibrium of (33.2) is asymptotically stable), then the maximum deviation of the motion from the equilibrium caused by crossing the boundary can be kept arbitrarily small if only the distance $|c|$ from the boundary is kept sufficiently small.

After the stability boundary has been crossed, the system oscillates without damping about the equilibrium. However, the amplitudes depend continuously on c and can be made arbitrarily small by suitable choice of c. Hence, in practice, the transition from the parameter value a to a + c with sufficiently small $|c|$ will, in general, have no undesired consequences. According to Bautin [1, 2], the stability boundary is said to have a *nondangerous domain* if the equilibrium is asymptotically stable at the points of the stability boundary which belong to this domain. A point at which the equilibrium is not asymptotically stable is said to belong to the *dangerous domain*. Both domains may be open or closed. In the neighborhood of a point of a dangerous domain oscillations of increasing amplitudes, irreversible transitions into other equilibria, or different other phenomena might occur. Geometrical interpretations of these situations were given by Malkin [19] and Hahn [4].

The theory of critical cases yields criteria for the dangerousness or non-dangerousness of a domain of the stability boundary. Lur'e [4, 7] established explicit criteria for the physical systems described by (14.1) and (14.3) under the restriction to the simplest critical cases treated in Sec. 30. The derivation of the criteria leads to the problem of expressing in closed form the constants g and γ_4, respectively, in terms of the parameters of the original equation. The signs of g and γ_4 are significant. Troickii [2] solved the corresponding problem for control systems with several controllers (Eq. (14.15)). However, the application of the criteria requires a considerable amount of computations, since the roots of the characteristic equation have to be known, etc. In practice, it is therefore convenient to use approximation procedures which, if sufficiently accurate, answer quickly the question of the "dangerousness" of a domain of the stability boundary (Aizerman [2], Magnus [1]).

The differential equation can also be written in the form (33.3) if a does not lie on the stability boundary. The effect of the quantity c always illustrates the dependence of the solutions on the parameters. To investigate this dependence, Dubošin [6] formally considered the quantity c as a function of t and added the equation

$$\dot{c} = 0$$

to the differential equation (33.3). The resulting system for the quantities $x_1, \ldots, x_n, c_1, \ldots, c_r$ can then be investigated by the direct method in the usual manner.

8 GENERALIZATIONS OF THE CONCEPT OF STABILITY

34. STABILITY IN A FINITE INTERVAL

According to Definition 2.1, stability in the sense of Liapunov pertains to the time interval $(0, \infty)$. Consequently, the stability property can never be verified in practice, since a physical system can be observed only during a finite time interval. Therefore, an attempt has been made to modify the concepts such that the motion has to be considered only during a finite time interval, and to develop corresponding criteria for *stability in a finite interval*.

Let the right hand side of the differential equation

$$\dot{x} = f(x, t) \tag{34.1}$$

belong to the class E for $|x| \leq h$, $t' \leq t \leq t''$. Furthermore, let $v(x, t)$ be a positive definite function and let $v(x, t) = $ const determine a cycle (cf. Sec. 3, Remark 2). By the *diameter* $d(t)$ of the domain

$$v(x, t) \leq a \tag{34.2}$$

bounded by the cycle we shall understand the least upper bound of the distance of any two points x_1, x_2 of (34.2) for fixed t. The function $d(t)$ is assumed to be bounded; a is a fixed positive number.

DEFINITION 34.1 (Lebedev [2]): Let $t' \leq t_0 \leq t \leq t_1 \leq t''$ and $d(t) \leq d(t_0)$. The equilibrium of (34.1) is said to be *stable in the interval* $[t_0, t_1]$ *with respect to the domain* (34.2), if the inequality

$$v(\mathsf{p}(t, x_0, t_0), t) \leq a \qquad (t_0 \leq t \leq t_1)$$

follows from $v(x_0, t_0) = a$. The equilibrium is called *unstable in the interval* $[t_0, t_1]$ *with respect to* (34.2) if this inequality is not valid.

SEC. 34 GENERALIZATIONS OF THE CONCEPT OF STABILITY

Lebedev introduced the concept of uniform stability with respect to (34.2) in the interval $[t', t'']$ if the equilibrium is stable in every interval $[t_0, t'']$ ($t' \leq t_0 \leq t''$). He called the stability monotonic if the integral curves always penetrate the cycle $v = \text{const}$ from the outside towards the inside (cf. also Gorbunov [1]).

To derive criteria (Kamenkov [3], Lebedev [1, 2], Kamenkov and Lebedev [1]), the original equation is written in a slightly different form:

$$\dot{x} = A(t) x + g(x, t) \tag{34.3}$$

In the interval $[t_0, t_1]$, the matrix $A(t)$ shall be continuous and bounded and it shall always have simple eigenvalues $\lambda_i(t)$. $A(t)$ need not necessarily be the matrix of the equation of the first approximation. Let $g(x, t)$ satisfy the conditions $g \in E$ and

$$|g| \leq c|x|^{1+\alpha} \quad (c > 0, \alpha > 0) \tag{34.4}$$

The nonsingular matrix $P(t)$ shall be chosen such that $PAP^I = D$ is a diagonal matrix. Since P is not yet uniquely determined by this condition, we may additionally require that P contain a constant element in each row and in each column. Finally, let the differentiable function $\varphi(t)$ be positive and bounded in the interval $[t_0, t_1]$. For the variable $y = \varphi(t) P(t) x$ a differential equation of the form

$$\dot{y} = Dy + My + g^*(y, t) \tag{34.5}$$

is obtained from (34.3), where

$$M = \dot{P}P^I + (\dot{\varphi}/\varphi) U \tag{34.6}$$

Now let

$$v(x, t) = |y|^2 = \sum_{i=1}^{n} |y_i|^2 = (\varphi(t))^2 x^T P^T \overline{P} x \tag{34.7}$$

If the smallest eigenvalue of the positive definite Hermitian matrix $P^T \overline{P}$ is denoted by $\gamma(t)$, then the diameter $d(t)$ satisfies the equation

$$\frac{d(t)}{d(t_0)} = \frac{\varphi(t_0)}{\varphi(t)} \sqrt{\frac{\gamma(t_0)}{\gamma(t)}} \tag{34.8}$$

(Lebedev [2]). From this it follows that $d(t)$ is bounded for suitable choice of $\varphi(t)$ and that $d(t)$ is monotonic as required by Definition 34.1. The total derivative of (34.7) for (34.5) has the form

$$\dot{v} = y^T (D + \overline{D} + M^T + \overline{M}) y + h(y, t) \tag{34.9}$$

where $|h| < c_1 |x|^{2+\alpha}$ as a consequence of (34.4). The function $\dot{v}$ is negative definite in the interval (t_0, t_1) if its first term is negative definite, i.e., if the Hermitian matrix

$$D + \overline{D} + M^T + \overline{M} = D + \overline{D} + (\dot{P}P^I)^T + \overline{\dot{P}P^I} + 2(\dot{\varphi}/\varphi) U \tag{34.10}$$

has only negative eigenvalues. This is the case if and only if the eigenvalues, taken with the opposite sign, of the matrix

$$H = \dot{P}P^I + (\dot{P}P^I)^T + D + \check{D} \tag{34.11}$$

are always larger than the function $d \log \varphi^2/dt$. This condition is also sufficient for stability of the equilibrium in the interval (t_0, t_1) with regard to the domain (34.2) defined by the function (34.7). If the condition is satisfied for $t = t_0$, the stability interval certainly extends to that value of t for which one of the eigenvalues of (34.11) violates the condition for the first time. (Lebedev [2] also derived conditions which assure stability in an interval as a consequence of stability in an adjacent interval.)

If the matrix (34.10) has a positive eigenvalue for $t = t_0$, then the equilibrium is unstable in each interval beginning at t_0. If the differential equation $\dot{x} = A(t) x$ is actually the equation of the first approximation for (34.1), then the previous considerations yield a sufficient condition for stability according to the first approximation in a finite interval: The eigenvalues of H, taken with the opposite sign, must always exceed the function $2(\dot{\varphi}/\varphi)$ by a fixed positive number δ.

If $A(t)$ has multiple eigenvalues at the initial instant t_0, the considerations must be modified in the following manner (Kamenkov [3], Lebedev [1]): The transformation matrix is chosen such that the blocks of the Jordan normal form $N(t)$ of $A(t)$ are at $t = t_0$ of the form

$$\begin{matrix} \lambda(t_0) & \frac{1}{2} & 0 & \cdots \\ \frac{1}{2} & \lambda(t_0) & \frac{1}{2} & \cdots \\ 0 & \frac{1}{2} & \lambda(t_0) & \cdots \\ \cdot & \cdot & \cdot & \cdots \end{matrix}$$

Then the differential equation for y is written in the form

$$\dot{y} = N(t_0)\, y + M^*y + h^*(y, t)$$

and the derivative of the Liapunov function (34.7) is formed. For small $|x|$ and in the neighborhood of t_0 the sign of $\dot{v}$ is determined by the behavior of the Hermitian form $y^T N(t_0)\, \bar{y}$. Thus, a necessary and sufficient condition for the existence of a stability interval is that this form be negative definite.

Letov [7, 8] used the concept of stability in a finite interval to treat Lur'e's problem (Sec. 14) in the case that the coefficients of the equations of motion (14.3) depend explicitly on t. As in the autonomous case, the problem is to determine conditions for the quantities $p_j = p_j(t)$ which guarantee stability in a finite interval for arbitrary initial points and arbitrary nonlinearities which are determined only by (14.2), or by (14.13). The function v entering the Stability Definition 34.1 is chosen as a quadratic form of all variables. As usual, a condition for absolute stability (Definition 14.1) is obtained by discussing the sign of its total derivative $\dot{v}$ for the equations of motion: A certain quadratic form of the quantities $p_j(t)$ and some auxiliary parameters, must be negative definite. However, the practical evaluation of this condition is rather difficult.

Štelik [1] and Czan [1, 3] established a relation between the concept of stability in a finite interval and the determination of optimal control (Sec. 6, Remark 6; cf. also Lebedev [4], S. K. Persidskii [2], Kudakova [1]).

35. DIFFERENTIAL EQUATIONS WITH BOUNDED SOLUTIONS

Applying the considerations of Sec. 3 to the case where the function v is positive in the exterior of a certain sphere and where $\dot{v}$ is negative in that domain, it can be concluded that the phase trajectories penetrate those hyper-surfaces $v =$ const which are sufficiently far away from the origin from the outside to the inside. Consequently, all solutions with bounded initial points are bounded themselves. If t is greater than a sufficiently large t-value, the absolute values of these solutions are even smaller than a fixed upper bound. Criteria for boundedness can be derived for special differential equations, cf. e.g., Reuter [1, 2]. Yoshizawa developed this method systematically in several publications ([2 to 8], in particular [9]). Apparently, his investigations are originally independent of the considerations of Liapunov. The analogy to the direct method is obvious and is emphasized by the fact that the boundedness of the solutions can be interpreted in the sense of a stability property of the trivial solution. LaSalle [2] introduced the terminology: "stability in the sense of Lagrange."

To the different types of stability (cf. Secs. 4 and 17) there correspond different types of boundedness. The most important types of them are defined in the following:

DEFINITION 35.1: The solutions of a differential equation are said to be *uniformly bounded* if there exists for a given $h > 0$ a constant $k > 0$ depending only on h such that

$$|p(t, x_0, t_0)| < k \quad \text{if} \quad |x_0| < h \quad (t \geq t_0)$$

DEFINITION 35.2: The solutions of a differential equation are said to be *ultimately bounded* with the bound b if there are positive numbers b and τ such that

$$|p(t, x_0, t_0)| < b \quad \text{if} \quad t > t_0 + \tau$$

The number τ might depend on x_0 and t_0.

DEFINITION 35.3: The solutions of a differential equation are said to be *uniformly ultimately bounded* with the bound b if the number τ of Definition 35.2 can be chosen independent of x_0 and t_0.

Yoshizawa defined some other types of boundedness corresponding to the concepts of "quasi-asymptotic stability," "equiasymptotic stability,"

130 GENERALIZATIONS OF THE CONCEPT OF STABILITY CHAP. 8

etc. Seibert [1] modified the definitions by assuming that the initial and final values of $|\mathsf{p}|$ are not located in spheres, but in more general domains.

To characterize the different types of boundedness by means of the direct method, "Liapunov functions" $v(\mathsf{x}, t)$ are introduced which have the properties explained in Sec. 1 in a domain

$$\mathfrak{L}_{h,t_0}: |\mathsf{x}| > h, \qquad t \geq t_0$$

In particular, the following properties are needed:

(a) in $\mathfrak{L}_{h,t_0}$, the function $v(\mathsf{x}, t)$ can be estimated above by a positive increasing function $\varphi_1(r)$:

$$v(\mathsf{x}, t) \leq \varphi_1(|\mathsf{x}|)$$

(b) in $\mathfrak{L}_{h,t_0}$, the function $v(\mathsf{x}, t)$ can be estimated below by a nonnegative increasing function $\varphi_2(r)$:

$$v(\mathsf{x}, t) \geq \varphi_2(|\mathsf{x}|)$$

(c) the derivative of v for the differential equation is nonpositive in $\mathfrak{L}_{h,t_0}$;

(d) the derivative of v for the differential equation can be estimated above in $\mathfrak{L}_{h,t_0}$ by a negative continuous function $-\varphi_3(r)$:

$$\dot{v} \leq -\varphi_3(|\mathsf{x}|)$$

The following theorems, for example, hold:

THEOREM 35.1: *The existence of a function $v(\mathsf{x}, t)$ with the properties (a), (b), and (c) is sufficient for uniform boundedness of the solutions.*

THEOREM 35.2: *The existence of a function $v(\mathsf{x}, t)$ with the properties (a), (b), and (d) is sufficient for uniform ultimate boundedness of the solutions.*

The conditions listed above are also necessary in the case where the right hand side of the differential equation belongs to the class C_1. It is possible to construct functions with these required properties.

Proof of Theorem 35.1: Let $r > h$. According to condition (a), the inequality

$$v(\mathsf{x}, t) \leq \varphi_1(r)$$

holds if $|\mathsf{x}| = r$. Because of condition (b), a number $r' > 0$ can be determined such that $\varphi_2(r') > \varphi_1(r)$. Suppose there would exist an initial point x_0 with $|\mathsf{x}_0| \leq r$ and a number t' such that

$$|\mathsf{p}(t', \mathsf{x}_0, t_0)| = r'$$

Since $|\mathsf{p}(t_0, \mathsf{x}_0, t_0)| = |\mathsf{x}_0| \leq r$ there must be two t-values t_1 and t_2 such that

$$t_0 \leq t_1 < t_2 \leq t'$$

$$|\mathsf{p}(t_1, \mathsf{x}_0, t_0)| = r, \qquad |\mathsf{p}(t_2, \mathsf{x}_0, t_0)| = r'$$

$$r < |\mathsf{p}(t, \mathsf{x}_0, t_0)| < r' \quad \text{if} \quad t_1 < t < t_2$$

If this were true, we would have
$$v(t_1) \leq \varphi_1(r), \quad v(t_2) \geq \varphi_2(r')$$
However, these inequalities contain a contradiction, since, according to condition (c), the function $v(t)$ does not increase. Consequently, $|\mathsf{p}(t, \mathsf{x}_0, t_0)| \leq r'$ if $|\mathsf{x}_0| \leq r$, i.e., the solutions are uniformly bounded.

Proof of Theorem 35.2: Proceeding in the same manner as in the proof of Theorem 35.1, we find that the solutions are uniformly bounded. Because of condition (b) we have $\varphi_2(r) \leq v(\mathsf{x}, t)$ if $|\mathsf{x}| > r$. Let $a > r$. Because of the uniform boundedness of the solutions there exists a number a' depending only on a such that
$$|\mathsf{p}(t, \mathsf{x}_0, t_0)| < a' \quad \text{if} \quad |\mathsf{x}_0| < a \quad \text{and if} \quad t \geq t_0$$
We consider the function $v(\mathsf{x}, t)$ in the domain
$$\mathfrak{G}: t \geq t_0, \quad r \leq |\mathsf{x}| \leq a'$$
According to property (d) there exists a number $-k$ depending only on a' such that in $\mathfrak{G}$ the inequalities
$$\dot{v}(t) \leq -k, \quad v(t) - v(t_0) \leq -k(t - t_0)$$
hold. In $\mathfrak{G}$ the function $v(t)$ has the following properties:
1. $v(t) > \varphi_2(r)$;
2. $v(t) \leq \varphi_1(a')$ according to property (a). Consequently, the trajectory must leave the domain $\mathfrak{G}$. However, this can happen only for decreasing $|\mathsf{x}|$. Thus, there is a number t' satisfying the relation
$$t_0 < t' < t_0 + (1/k)(\varphi_1(a') - \varphi_2(r))$$
such that
$$|\mathsf{p}(t', \mathsf{x}_0, t_0)| = r$$
This is the condition for uniform ultimate boundedness.

The analogy between these proofs and the proofs of the corresponding theorems of the theory of stability is obvious.

Yoshizawa [4, 9] defined also the analogue to total stability which characterizes the effect of perturbation terms on the boundedness behavior. By means of Liapunov functions, Yoshizawa established sufficient conditions.

An important application of the theorems on boundedness can be found in the proof of the existence of periodic solutions, in particular of systems of the order 2 (cf. e.g., Kuškov [1], Zheleznov [1], Yoshizawa [10]).

Reissig [3, 4] pointed out that ultimate boundedness of the solutions can be interpreted as complete instability (cf. Definition 2.6) of the point at infinity. To obtain a criterion, we have to transform that one formulated in the Corollary of Theorem 5.2, so to speak, to the point at infinity.

132 GENERALIZATIONS OF THE CONCEPT OF STABILITY   CHAP. 8

The definitions of the different types of boundedness as well as the definitions of the different types of stability refer to the trivial solution. The quantity $|p(t, x_0, t_0)|$ occurring in these definitions represents the distance of the motion from the origin. Yoshizawa [9, 11] considered also systems for which the distance between corresponding points of any two solutions is uniformly bounded, or ultimately bounded, or tending towards zero, etc. For example, Yoshizawa called the *system* $\dot{x} = f(x, t)$ *equi-distance-bounded* if there exists a positive function $\varphi(r, t)$ such that

$$|p(t, x_0, t_0) - p(t, x_0', t_0)| \leq \varphi(t_0, \alpha)$$

provided that

$$|x_0 - x_0'| \leq \alpha$$

If φ tends monotonically towards 0 with decreasing r, the system is said to be *equistable*. It is said to be *uniform-distance-bounded* if φ can be chosen independent of t, etc. The remarkable point is that all these different types of behavior can be characterized by means of Liapunov functions v which depend, however, on two vector arguments, $v = v(x, y, t)$. The vectors x and y are interpreted as solutions of

$$\dot{x} = f(x, t) \quad \text{and} \quad \dot{y} = f(y, t)$$

with the same function f. The statements are based on the time derivative $\dot{v}$. For example, the following THEOREM holds: The system is equistable if there exists a function $v(x, y, t)$ such that the inequalities

$$\varphi_1(|x - y|) \leq v(x, y, t) \leq \varphi_2(|x - y|)$$

$$\dot{v}(t) \leq 0$$

hold, where $\varphi_1(r)$, $\varphi_2(r)$ are suitable functions of the class K.

In most of his definitions and proofs, Yoshizawa did not require uniqueness of the solutions.

36. THE APPLICATION OF THE DIRECT METHOD IN GENERAL METRIC SPACES

A considerable portion of the theory of the direct method developed in the previous chapters is not restricted to motions which are determined by differential equations. Certain portions of this theory can be generalized in two directions. One way of generalizing the theory has already been indicated in Definition 2.1 and in Sec. 20: the motions considered there do not necessarily have to be integral curves of differential equations. A second generalization, which is quite independent of the first, is to discard the restriction to a finite dimension. The "point" describing the motion does not have to be an element of the n-dimensional Euclidian phase space or of the $(n + 1)$-dimensional motion space. Going over to infinitely many dimensions it is possible to establish a theory of stability in general metric spaces, as (according to Zubov [6]) will be sketched in the following. (With regard to the theory of metric spaces cf., e.g., Kolmogorov and Fomin [1].)

Let x be an element of the metric space R and let $\rho(x)$ be the norm of x. The parameter t is restricted to nonnegative real numbers. By $F_{t_0}^t$ we denote a mapping, not necessarily one-valued, which depends on two parameters t_0 and $t \geq t_0$, and which associates with the element x the set $F_{t_0}^t(x) \subset R$.

DEFINITION 36.1: A two-parameter family of mappings of the space R onto itself is called a *complete general system* if the following conditions are satisfied:

1. The nonempty set $F_{t_0}^t(x) \subset R$ is defined for each $x \in R$, for each $t_0 \geq 0$ and for each $t \geq t_0$.
2. $\lim_{t \to t_0 + 0} F_{t_0}^t(x) = x$.
3. If $t_1 > t_0$ and if x_1 is any element of the set $F_{t_0}^{t_1}(x)$, then

$$\bigcup F_{t_1}^t(x_1) = F_{t_0}^t(x) \qquad (t \geq t_1, x_1 \in F_{t_0}^{t_1}(x))$$

If condition 1 is weakened by the assumption that the mapping is defined only for a subset $R_1 \subset R$, then the system is called an *incomplete general system*.

The union of the sets $F_{t_0}^t(x)$ for $t \geq t_0$ forms a *motion*. A set $\mathfrak{M} \subset R$ consisting of motions is called an *invariant set* of the general system. The *distance* of a point x from a set $\mathfrak{M}$ is defined by

$$\rho(x, \mathfrak{M}) = \inf \rho(x, y) \qquad (y \in \mathfrak{M})$$

A *neighborhood* $\mathfrak{U}(\mathfrak{M}, r)$ of $\mathfrak{M}$ is the union of all points x for which $0 < \rho(x, \mathfrak{M}) < r$. Consequently, $\mathfrak{M}$ does not belong to $\mathfrak{U}(\mathfrak{M}, r)$; cf. Sec. 20. The distance of the set $F_{t_0}^t(x_0)$ from $\mathfrak{M}$ is defined by

$$\rho(x_0, t, t_0; \mathfrak{M}) = \sup \rho(x, \mathfrak{M}) \qquad (x \in F_{t_0}^t(x_0)) \qquad (36.1)$$

For fixed $\mathfrak{M}$, this distance is a function of x_0 and the parameters t and t_0.

If the definition of the dynamical system in the Euclidean space (Sec. 20) is transferred to the metric space R by replacing the distance $|x|$ by the norm $\rho(x)$, the concept of the dynamic system in the metric space is obtained. This concept is a special case of the concept of the general system.

DEFINITION 36.2: The invariant set $\mathfrak{M}$ of a general system is called *stable* (in the sense of the metric of the space R) if for each $\epsilon > 0$ there exists a number $\delta > 0$ such that

$$\rho(x_0, t, t_0; \mathfrak{M}) < \epsilon \qquad (t \geq t_0 \geq 0)$$

provided that

$$\rho(x_0, \mathfrak{M}) < \delta$$

The other basic Definitions (2.2, 2.3, 2.5, 17.1, 17.3, Remark 7 of Sec. 17) can be transferred analogously.

A *functional* $v(x, t)$ depending on the parameter t takes the place of the Liapunov function. Instead with the time function $v(t)$ as defined in Sec. 1(i) (which represents the Liapunov function along a motion), we have to operate now with

$$v(x, t, t_0) = \sup v(y, t) \qquad (y \in F^t_{t_0}(x)) \tag{36.2}$$

To characterize the stability properties of an invariant set $\mathfrak{M}$, we consider a real functional with the following properties:

(a) $v(x, t)$ is defined for all $t \geq 0$ and for all x of a certain neighborhood $\mathfrak{U}(\mathfrak{M}, r)$ of $\mathfrak{M}$;

(b) for each sufficiently small $\eta_1 > 0$ there exists a number $\eta_2 > 0$ such that $v(x, t) > \eta_2$ for all $t \geq 0$, provided $\rho(x, \mathfrak{M}) > \eta_1$;

(c) $\lim v(x, t) = 0$ uniformly with regard to $t \geq t_0$ if $\rho(x, \mathfrak{M}) \to 0$;

(d) the function $v(x, t, t_0)$ defined by (36.2) does not increase for $t \geq t_0$;

(e) the function $v(x, t, t_0)$ tends towards zero if $t \to \infty$ and if x belongs to a certain neighborhood $\mathfrak{U}(\mathfrak{M}, \delta)$ of $\mathfrak{M}$;

(f) $\lim\limits_{t-t_0 \to \infty} v(x, t, t_0) = 0$ uniformly with regard to t_0 if $\rho(x, \mathfrak{M}) \leq \delta$.

Conditions (a) through (d) are necessary and sufficient for stability, (a) through (e) for asymptotic stability and (a) through (f) for uniform asymptotic stability of the invariant set $\mathfrak{M}$ (Zubov [6]).

An even greater analogy to the theorems of the direct method is represented by:

THEOREM 36.1 (Zubov [6]): The following conditions are necessary and sufficient for the invariant set $\mathfrak{M}$ of the general system to be uniformly asymptotically stable and uniformly attracting:

1. There exist two functionals $v(x, t)$ and $w(x, t)$ with the properties (a) through (c).

2. The associated functions $v(x, t, t_0)$ and $w(x, t, t_0)$ defined by (36.2) satisfy the relation

$$\frac{d}{dt} v(x, t, t_0) = -w(x, t, t_0)$$

A corresponding theorem holds for instability.

If more details are known about the general system, statements can be made which are more far reaching. For example, the construction method of Sec. 21 can be transferred to dynamical systems of a metric space. Zubov proved furthermore that the function (36.1) can be estimated by means of the functional $v(x, t)$ introduced in Theorem 36.1.

Many of the results of Chaps. 2 and 4 which have been derived by a direct approach can be obtained by specialization from the elegant theorems on the theory of the general systems. However, it has to be realized that the derivation of stability criteria in a strict sense is not actually the goal of the theory of general or dynamic systems. Of primary interest is the characterization of the motions of the system. Therefore, the direct proof of a stability criterion is often simpler than its derivation from general topological theorems.

Important special cases of general systems are represented by the trajectories of differential equations

$$\dot{x} = f(x, t) \qquad (f(0, t) \equiv 0) \tag{36.3}$$

in linear metric spaces, in particular in Banach spaces. Here, f has to be considered as an operator (depending on a parameter t) to be applied on the element x of the space (cf. Kolmogorov and Fomin [1]). In order to apply the direct method to such differential equations without referring to the theory of the general systems, it is necessary to find out which of the theorems of Chaps. 2 through 5 are, with regard to their proofs, independent of the number of dimensions. By extending the basic definitions correspondingly, these theorems remain valid for differential equations in general spaces. In this manner, Massera [3, 4] found that the following theorems, for example, hold in Banach spaces: Theorems 4.1, 4.2, 4.3, and 5.1 (sufficient conditions for stability, asymptotic stability, asymptotic stability in the whole, and instability of the equilibrium; the theorem on instability remains valid even in a more general form given by Massera [4]), Theorem 17.6 (sufficient condition for uniform asymptotic stability), the Converse Theorems 18.3 and 18.4 with the somewhat weaker statement $v \in C_0$ (Massera's proof for the differentiability of v up to any order essentially makes use of the finiteness of the number of dimensions); under the assumption that $f \in C_s$, there exists a functional v with $v \in C_s$ (Massera [3]). Corollary 2 of Theorem 18.3 as well as Theorem 24.3 remains also valid. In all these theorems, the "total derivative of the functional $v(x, t)$ for the differential equation (36.3)" has either to be replaced by Frechet's differentiation process

$$\dot{v} = \lim_{h \to 0} \frac{v(x + hf, t + h) - v(x, t)}{h}$$

or to be defined by the upper limit of the difference quotient for $h \to 0+$ (cf. Krasovskii [16]).

Some theorems cannot be transferred. For example, Theorem 17.5 (equivalence of asymptotic stability and uniform asymptotic stability in the case of autonomous differential equations) is not valid in general spaces. An example is given by the differential equation

$$\dot{x}_n = -\frac{1}{n} x_n \qquad (n = 1, 2, \ldots) \tag{36.4}$$

in the Hilbert space (Massera [4]). The proof of Theorem 17.5 collapses since the function $\rho(p(t, x_0, t_0))$ cannot be estimated uniformly by a monotonically decreasing function of $t - t_0$ for $\rho(x_0) \leq r$. Theorem 19.4 (necessary condition for complete instability) also is not valid in general spaces (S. K. Persidskii [1]). Furthermore, the considerations which are connected with Sec. 8 cannot be transferred, since there are no analogues to Theorems 8.1 and 8.2 on the stability behavior of the equilibrium of autonomous differential equations.

The first extensions of the direct method to spaces of a more general type were accomplished by K. P. Persidskii [6, 7]. He considered (36.3) as a system of countably many differential equations for the components $x_1, x_2, \ldots$ of the vector x with the norm

$$\rho(x) = \sup\,(|x_1|, |x_2|, \ldots)$$

In the so defined space, a stability theorem holds which has no analogue in the Euclidean space:

THEOREM 36.2 (K. P. Persidskii [7]): *The equilibrium is stable if the following conditions are satisfied. There exists a continuous positive semi-definite functional $v(x, t) \in C_1$ with the properties:*

(a) $v(x, t) \leq \alpha$ *implies* $|x_1| \leq g(\alpha)$, *where $g(r)$ is an arbitrary, but fixed, positive, continuous function of r with $g(0) = 0$;*

(b) *for each value $s = 1, 2, \ldots$ there exists a sequence $x_s, x_{s1}, x_{s2}, \ldots$ of components of the solution vector such that the total derivative $\dot{v}(x, t)$ for (36.1) is negative semi-definite for $x = \{x_s, x_{s1}, \ldots\}$. The equilibrium is asymptotically stable if $\dot{v}(x, t)$ is negative definite.*

Furthermore, K. Persidskii [7, 8, 12, 13] transferred Theorems 18.1, 18.2, 19.3, the converse of Theorem 5.3 and Theorem 26.2 (stability according to the first approximation) to systems of countably many differential equations. However, to transfer Theorem 26.2, he did not apply the direct method. Charasachal [1] gave a proof in which a Liapunov function is used. Goršin [1, 3] transferred the concept of total stability.

Under very general assumptions, Massera and Schäffer [1 to 4] investigated the relations between the following problems:

1. Existence of bounded solutions of the nonhomogeneous linear differential equation.

2. Boundedness, or exponential decay, respectively, of the solutions of the corresponding homogeneous equation.

3. Stability according to the first approximation.

4. Existence of Liapunov functions.

Movchan [2] noted that the definition of stability in general metric spaces is actually based on two different metrics: the metric in the space of the "initial points" and the metric in the motion space. If the initial points and the motions are interpreted in two different spaces, these two metrics need not be identical, although they are not completely independent of each other. Movchan developed this idea and established a theory of stability with two metrics where certain functionals replace the Liapunov functions.

37. STABILITY IN THE CASE OF PARTIAL DIFFERENTIAL EQUATIONS

Let
$$\partial u_s/\partial t = f_s(x_1, \ldots, x_k; u_1, \ldots, u_n; \ldots, \partial u_i/\partial x_j, \ldots) \quad (37.1)$$
$$(s = 1, \ldots, n; i = 1, \ldots, n; j = 1, \ldots, k)$$

be a system of partial differential equations of the first order. The right hand sides shall be continuous in a certain domain of their arguments. Let R be a metric space, whose elements are function vectors $z(x)$ with the components $z_1(x), \ldots, z_n(x)$. With every z there shall be associated a solution $u = u(t, z)$ of (37.1). Each of these solutions shall be defined as an element of R for all t ($-\infty < t < +\infty$) and, in particular, $u(t, z)$ shall be a continuous functional of its arguments and shall satisfy the conditions $u(0, z) = z$ and $u(t, 0) \equiv 0$. The so defined solutions of the system (37.1) define in R a dynamical system for which the solution $u = 0$ is an invariant set (cf. Sec. 20). The solution $u = 0$ is called stable (in the sense of the metric of R), if for every $\epsilon > 0$ there exists a number $\delta > 0$ such that the inequality $\rho(u(t, z), 0) < \epsilon$ for $t \geq 0$ follows from $\rho(z, 0) < \delta$. The concepts of asymptotic stability, etc., are defined correspondingly.

If the independent variable t occurs explicitly in the right hand sides of the differential equations, then the solution depends not only on t and z, but also on the "initial instant" t_0. Consequently, the notation $u = u(t, z, t_0)$ with $u(t_0, z, t_0) = z$ has to be used. The solutions, which are now defined only for $t \geq t_0$, do not form a dynamical system; they form a more general system in R which also contains $u = 0$ as an invariant set. The stability theory of dynamical, or general systems, therefore, can be applied directly. The detailed formulation of the general theorems for the dynamical case was given by Zubov [6].

In place of (37.1) let the more special system
$$\frac{\partial u_s}{\partial t} = f_s(u_1, \ldots, u_n) + \sum_{i=1}^{k} b_i \frac{\partial u_s}{\partial x_i} \quad (s = 1, \ldots, n) \quad (37.2)$$

with constant coefficients $b_i (i = 1, \ldots, k)$ and the system of ordinary differential equations
$$du_s/dt = f_s(u_1, \ldots, u_n) \quad (s = 1, \ldots, n) \quad (37.3)$$

be given. The functions f_s are continuous and bounded and they vanish at the origin. Denoting the general solution of (37.3) by $p(t, u_0, t_0)$ (cf. Sec. 1), the solution of (37.2), with the initial points $z(x)$ is obtained in the form
$$p(t, z(x + tb), 0) \quad (37.4)$$

(Zubov [6]). The constants $b_1, \ldots, b_n$ are the components of the vector b. The relation (37.4) permits a comparison of the stability behavior of the trivial solutions of the differential equations (37.2) and (37.3). If the equilibrium of (37.3) is asymptotically stable, two functionals can be constructed by means of the functions $v(x)$ and $w(x)$ whose existence is guaranteed by Theorem 20.1. These functionals satisfy the assumptions for asymptotic stability of the partial differential equation (37.2). The converse is also valid (Zubov [6]). Consequently, the partial differential equation (37.2) has an asymptotically stable equilibrium if the same is true for (37.3), and vice versa. In a certain sense, the theorem on stability according to the first approximation holds for (37.2).

Zubov [6] proved a similar theorem which relates the stability behavior of the equilibrium of a system of partial differential equations of higher order, namely

$$\frac{\partial u_s}{\partial t} = \sum_{i=1}^{n} \sum_{\alpha_i} a_{si}^{(\alpha_1, \ldots, \alpha_k)}(t) \frac{\partial^{\alpha_1 + \cdots + \alpha_k} u_i}{\partial x_1^{\alpha_1} \cdots \partial x_n^{\alpha_k}}$$

$$(s = 1, \ldots, n; 0 \leq \alpha_1 + \cdots + \alpha_k \leq m)$$

to the stability behavior of the equilibrium of the system

$$\frac{du_s}{dt} = \sum_{i=1}^{n} a_{si}^{(0, \ldots, 0)} u_i \qquad (s = 1, \ldots, n)$$

of ordinary differential equations.

Movchan [1] applied the direct method to the special equation

$$\frac{\partial^4 w}{\partial x^4} - a\frac{\partial^2 w}{\partial x^2} + \frac{\partial^2 w}{\partial t^2} = 0 \qquad (w = \partial^2 w/\partial x^2 = 0 \text{ if } x = 0, \text{ or } x = 1)$$

He obtained in this special case the general theorems proved by Zubov [6] in a simpler manner without using the theory of dynamic systems. The essential point is to define the metric in the space of the solutions in a manner which suits the special problem. By means of the Liapunov functional

$$\int_0^1 (w_{xx}^2 + aw_x^2 + w_t^2)\, dx$$

Movchan obtained the value for Euler's first (critical) load which is known from the theory of vibrations of plates. The investigation shows that the method might lead to practical results in concrete problems.

Volkov [1] considered a certain family of solutions from among the set of all solutions of a hyperbolic differential equation which is determined, for instance, by the condition that the initial points belong to a certain set of functions. He then considered an operator J which associates to the solution $u(x, t)$ a functional $J(u)$ depending on t. For J, or $J(u)$, respectively, the concept "definite" is defined by means of an integral inequality.

The stability of the trivial solution $u = 0$ is defined by means of the functional: The trivial solution is said to be stable, or asymptotically stable, respectively, if $dJ/dt \leq 0$ or $\lim J(u) = 0$ with $t \to \infty$, respectively. Further investigations (e.g., independence of the stability on particular functionals) have not yet been carried out.

38. APPLICATION OF THE DIRECT METHOD TO DIFFERENTIAL-DIFFERENCE EQUATIONS

Let the system

$$\dot{x}_i(t) = f_i(x_1(t - h_{11}), \ldots, x_1(t - h_{m1}), \ldots, x_j(t - h_{kj}), \ldots; t)$$

$$(i = 1, \ldots, n; j = 1, \ldots, n; k = 1, \ldots, m) \quad (38.1)$$

of differential-difference equations be given. The functions f_i shall be defined for (x, t)-values of the domain $\mathfrak{R}_{a,0}$ of the motion space, and, as functions of their arguments

$$y_{ik} = x_i(t - h_{ki}) \quad (i = 1, \ldots, n; k = 1, \ldots, m) \quad (38.2)$$

they shall satisfy the Lipschitz condition

$$|f_i(y_1, t) - f_i(y_2, t)| < l|y_1 - y_2| \quad (38.3)$$

uniformly in $\mathfrak{R}_{a,0}$. (y is the vector of the quantities y_{ik}; the subscripts 1 and 2 denote different y-vectors.) Furthermore, $f_i(0, t) \equiv O$ is assumed. Differential-difference equations are characterized by the *delay terms* $h_{ki}(t)$. Let the delay terms satisfy the inequalities

$$0 \leq h_{ki}(t) \leq h_i \leq h \quad (t \geq 0) \quad (38.4)$$

and let these functions be piecewise continuous. Constant delay terms are of particular interest for practical applications.

A solution of (38.1) for $t > t_0 \geq h$ is uniquely determined by the *initial points*

$$\lim_{t \to t_0 + 0} x_i(t) = x_{i0} \quad (38.5)$$

and by the *initial functions*

$$x_i(t) = z_i(t) \quad (t_0 - h_i \leq t \leq t_0) \quad (38.6)$$

(Myškis [1], Hahn [1]). The initial functions, considered as components of a vector $z(t)$, are assumed to be continuous and bounded. In addition, let

$$\lim_{t \to t_0 + 0} z_i(t) = x_{i0}$$

(The last assumption is not essential; however, it leads to some simplifications.)

In analogy to the notation of Sec. 1, the solutions of the differential-difference equation, depending on the arbitrary initial vector $z(t)$, shall be denoted by

$$\mathsf{p}(t, \mathsf{z}(t_0), t_0) \tag{38.7}$$

The initial vector is defined in the interval $(t_0 - h, t_0)$. A solution, in other words, a motion, is obtained by "continuation" of $z(t)$ beyond this interval. It can be considered as an element of a metric space with the norm

$$\rho(\mathsf{x}(t)) = \sup |x_i(t - \tau)| \quad (0 \le \tau \le h; i = 1, \ldots, n) \tag{38.8}$$

where t has to be considered as a parameter. In a similar manner as a solution of a differential equation, the solution (38.7) represents a two-parameter mapping of the space of the norm (38.8) onto itself. However, it does not define a general system in the sense of Definition 36.1, since Property 3 is not necessarily satisfied. Not all of its trajectories initiating at $t_1 > t_0$ can be considered as continuations of solutions initiating at t_0. Thus, the stability property mentioned in Sec. 2, Remark 1, is lost: it can happen that a particular solution is stable with regard to all initial functions defined in the interval $(t_0 - h, t_0)$, whereas the same solution is unstable if referred to the initial instant $t_1 > t_0$. (Examples of this type were given by Elsgolts [2] and Malkin [16].) As a consequence, Elsgolts [1, 2] included the dependence on the initial instant in the definition of stability.

DEFINITION 38.1: The solution (38.7) is said to be *stable* if for every $\epsilon > 0$ and every $t_1 > t_0$ there is a number $\delta > 0$ depending on t_1 and ϵ (and t_0) with the following property: The inequality

$$\rho[\mathsf{p}(t, \tilde{\mathsf{z}}(t_1), t_1) - \mathsf{p}(t, \mathsf{z}(t_0), t_0)] < \epsilon \quad (t \ge t_1 \ge t_0)$$

is satisfied for every function $\tilde{\mathsf{z}}(t)$ which satisfies the inequality

$$\rho[\tilde{\mathsf{z}}(t_1) - \mathsf{p}(t_1, \mathsf{z}(t_0), t_0)] < \delta$$

The definition can be simplified by considering the equilibrium instead of the solution (38.7). This leads to:

DEFINITION 38.2: The equilibrium of (38.1) is said to be *stable* if for every $\epsilon > 0$ and for every $t_1 \ge t_0 \ge h$ there is a number $\delta(\epsilon, t_1)$ such that

$$\rho(\mathsf{p}(t, \mathsf{z}(t_1), t_1)) < \epsilon \quad (t > t_1)$$

if

$$\rho(\mathsf{z}(t_1)) < \delta$$

The concepts of uniform stability, asymptotic stability, uniform asymptotic stability, and exponential stability can be defined accordingly. Because of the reasons mentioned above, it is not possible to obtain the stability theory for differential equations by specializing the stability theory for general systems, as is possible in the case of ordinary differential equations. In spite of this, the stability theories of the two types of equations are quite similar. Elsgolts [1, 2] showed that Theorems 4.1, 4.2, and 5.1 can be transferred verbatim. However, the total derivative

$$\dot{v} = \sum_{i=1}^{n} \frac{\partial v}{\partial x_i} f_i + \frac{\partial v}{\partial t}$$

of the Liapunov function v for (38.1) depends not only on the arguments $x_1, \ldots, x_n, t$, but also on the quantities (38.2). Therefore, in order to determine the sign of $\dot{v}$, we have to know either the quantities y_{ik} along a motion, which contradicts the idea of the direct method, or we have to know that $\dot{v}$ is definite as a function of the vectors x and y, which is a very strong requirement. The assumptions of Theorem 4.2 are, as a matter of fact, not feasible even in the case of very simple differential-difference equations with $n = 1$, whose solutions approach zero oscillatingly. Theorem 4.2 already assures asymptotic stability in the sense of the Euclidean norm of the solution, and, as a consequence of $n = 1$, even the monotonic approach of $x(t)$ towards 0.

Krasovskii [17] developed the stability theory for differential-difference equations by means of Liapunov *functionals*. This is a more suitable approach. The functionals are defined in the space of the initial functions and they depend on the parameter t. Let these functionals be denoted by $v(\mathsf{z}(-\tau), t)$. The auxiliary variable in the argument of z_i varies in the interval $0 \leq \tau \leq h_i$. The functional v shall be defined in a domain

$$\mathfrak{K}_{b,t_0}: \rho(\mathsf{z}) \leq b, \quad t \geq t_0$$

of its arguments and shall be continuous with respect to the norm (38.8). In analogy to (1.8), we introduce the concepts "definiteness" and "decrescentness" of a functional. For example: the functional v is said to be *definite* if

$$v(\mathsf{z}(-\tau), t) > \varphi(\rho(\mathsf{z}(0))) \quad (t \geq t_0)$$

where φ belongs to the class K, etc. Based on these definitions, we have:

THEOREM 38.1 (Krasovskii [17]): If a positive definite decrescent functional v exists such that its total derivative for (38.1)

$$\dot{v} = \frac{d}{dt} v(\mathsf{p}(t - \tau, \mathsf{z}(t_0), t_0), t)$$

is a negative definite functional, then the equilibrium is uniformly asymptotically stable.

Like Theorem 4.2, this theorem permits several generalizations. For instance, the expression which is analogous to (1.11) can be used in place of $\dot{v}$. The condition "$\dot{v}$ negative definite" can be replaced by $\dot{v} < -\varphi_1(\mathsf{x}(t))$, where $\mathsf{x}(t)$ represents the solution and φ_1 a positive definite function of the arguments $x_1, \ldots, x_n$. A modified proof was given by Halanaj [1].

The proofs of these statements are similar to those of the corresponding theorems for ordinary differential equations. This is also the case for the converse of Theorem 38.1 which is based, as well as the proof of Theorem 19.3, on the method of Massera [1].

THEOREM 38.2 (Krasovskii [17]): If the equilibrium of (38.1) is uniformly asymptotically stable, there exists a functional v which satisfies the assumptions of Theorem 38.1 and which, in addition, satisfies a Lipschitz condition

$$|v(z_1(-\tau), t) - v(z_2(-\tau), t)| < l_1 \rho(z_1(0) - z_2(0))$$

If perturbation terms are added to the right hand side of (38.1), systems of equations are obtained which suggest the definition of total stability for differential-difference equations. This condition was first stated by Elsgolts [1, 2] in a form corresponding to Definition 28.1; Krasovskii [17] generalized this definition somewhat.

DEFINITION 38.3: In addition to (38.1), let the "perturbed" system

$$\dot{x}_i = f_i(x_1(t - h^*_{11}), \ldots, x_n(t - h^*_{mn}); t)$$
$$+ g_i(x_1(t - h^*_{11}), \ldots, x_n(t - h^*_{mn}); t) \qquad (38.9)$$

be given. The equilibrium is said to be *totally stable*, if for every $\epsilon > 0$ three numbers $\delta > 0$, $\eta_1 > 0$ and $\eta_2 > 0$ can be so determined, that for the solution $\mathsf{p}(t, \mathsf{z}(t_0), t_0)$ of the perturbed differential-difference equations (38.9) the inequality

$$\rho[\mathsf{p}(t, \mathsf{z}(t_0), t_0)] < \epsilon \qquad (t \geqq t_0)$$

is valid, provided the conditions

$$\rho[\mathsf{z}(t_0)] < \delta, \qquad |g_i(\mathsf{x})| < \eta_1 \quad \text{in} \quad \Re_{\epsilon, t_0}$$
$$|h_{jk}(t) - h^*_{kj}(t)| < \eta_2, \qquad h^*_{jk}(t) \geqq 0$$

are satisfied.

The following theorem holds for total stability of differential-difference equations in the sense of Definition 38.3 (and also for its generalization defined by (28.5)).

THEOREM 38.3 (Krasovskii [17], Germaidze and Krasovskii [1]): Uniform asymptotic stability implies total stability.

Sufficient conditions for integral stability are given by Halanaj [4]. Theorem 38.3 contains as a special case the following theorem which was proved earlier by Barbašin and Krasovskii [1] (cf. also Krasovskii [16]) in connection with the direct method for ordinary differential equations.

THEOREM 38.4: If the equilibrium of the ordinary differential equations

$$\dot{x}_i = f_i(x_1, \ldots, x_n, t)$$

is uniformly asymptotically stable, then the same is true for the equilibrium of the differential-difference equations

$$\dot{x}_i = f_i(x_1(t - h_{i1}), \ldots, x_n(t - h_{in}); t)$$

provided the delay terms h_{ij} are sufficiently small.

Of particular interest is the case where the unperturbed system is linear. Let (38.9) be of the form

$$\begin{aligned}\dot{x}_i &= p_{i1}(t)\, x_1(t - h_{i1}) + \cdots + p_{in}(t)\, x_n(t - h_{in}) \\ &\quad + g_i(\ldots, x_j(t - h_{jk}^*), \ldots; t) \\ &\quad (i = 1, \ldots, n;\, j = 1, \ldots, n;\, k = 1, \ldots, m)\end{aligned} \quad (38.10)$$

Here, the $p_{ij}(t)$ are continuous bounded functions which satisfy a Lipschitz condition

$$|p_{ij}(t') - p_{ij}(t'')| < a|t' - t''| \quad (38.11)$$

The delay terms h_{ij}, h_{ij}^* are positive constants. For the functions g_i, estimates of the form

$$|g_i(x_1, \ldots, x_n, t)| < a_1(|x_1| + \cdots + |x_n|)^{1+\gamma} \quad (\gamma > 0)$$

hold. In addition to (38.10), the autonomous system

$$\dot{y}_i(t) = p_{i1}(\beta)\, y_1(t - h_0) + \cdots + p_{in}(\beta)\, y_n(t - h_0) \quad (38.12)$$

is considered. The number β is to be regarded as a parameter, and for the constant delay term h_0, the inequalities

$$|h_{ij} - h_0| < k, \qquad |h_{ij}^* - h_0| < k \quad (38.13)$$

hold for fixed k. Under these assumptions the following theorem is valid:

THEOREM 38.5 (Krasovskii [17]): Let the equilibrium of (38.12) be exponentially stable, i.e., let

$$\rho[\mathsf{p}(t, \mathsf{z}(t_0), t_0)] < b\rho(\mathsf{z}(t_0))\, e^{-\alpha(t-t_0)} \quad (38.14)$$

for all $\beta > 0$ with fixed numbers $b > 0$, $\alpha > 0$. The numbers a and k of (38.11), and (38.13), respectively, can then be so chosen that the equilibrium of (38.10) is asymptotically stable.

In the case of exponential stability (cf. (38.14)) the theorems for the first approximation hold in a similar sense as for ordinary differential equations (Germaidze [2], Halanaj [1 to 4]). In this connection, Halanaj [1] proved in analogy to Theorem 24.3: If the equilibrium of a linear differential-difference equation

$$\dot{\mathsf{x}}(t) = \mathsf{A}_0(t)\,\mathsf{x}(t) + \mathsf{A}_1(t)\,\mathsf{x}(t - h_1) + \cdots + \mathsf{A}_k(t)\,\mathsf{x}(t - h_k)$$

is uniformly asymptotically stable, then it is also exponentially stable.

To prove this theorem, the operator of the mapping determined by (38.7) is used which transforms z into p. This operator corresponds to the operator Z of the proof of Theorem 24.3. Since the operator determined by (38.7) is linear and bounded, the reasoning applied to Z can be repeated.

If the coefficients and the delay terms of the linear system are periodic functions with the same period and if the equilibrium of this system is exponentially stable, then a "sufficiently adjacent" nonlinear system with periodic coefficients and delay terms has an asymptotically stable periodic solution (Krasovskii [22]).

A sufficient condition for exponential stability of the equilibrium of (38.12) is that the "characteristic equation"

$$\det (p_{ij}(\beta) - \lambda e^{\lambda h_0} \cup) = 0$$

has only roots $\lambda_1, \lambda_2, \ldots$, whose real parts are smaller than a fixed negative number λ_0 (Wright [1]). The assumption of Theorem 38.5 is satisfied if the number $\lambda_0 = \lambda_0(\beta)$ is larger than a positive number which does not depend on β.

If the differential-difference equations are linear, if they have constant coefficients and constant delay terms, and if the equilibrium is exponentially stable (similarly as in Sec. 22), the functional can be chosen as an integral of the form

$$\int_t^\infty \rho^2(\mathsf{p}(\tau, \mathsf{z}(t), t) \, d\tau$$

In this case it is convenient to define the norm by

$$\rho(\mathsf{x}(t)) = \sqrt{|\mathsf{x}(t)|^2 + \int_0^h |\mathsf{x}(t - \tau)|^2 \, d\tau}$$

or some similar expression, rather than by (38.8). Since in the present case the general solution of the differential-difference equation can be written explicitly, it is also possible to give the functional v explicitly, either as an integral expression or as a quadratic form in infinitely many variables; here, the variables are the quantities (38.5) and, for instance, the Fourier coefficients of the functions (38.6).

The application of Theorem 38.1 is limited by the fact that suitable functionals v can presently be given only for a few special differential-difference equations. Razumichin [4] attempted to make Krasovskii's theorems somewhat more applicable. Instead of the functional, he considered a Liapunov function $v(\mathsf{x}, t)$, defined in the motion space, and he used the derivative $\dot{v}$ as a functional. This functional $\dot{v}$ does not necessarily have to be negative definite; it is sufficient that $\dot{v}$ be negative definite along certain solutions of (38.1). For instance, the following theorem holds:

THEOREM 38.6 (Razumichin [4]): *The equilibrium of* (38.1) *is stable if there exists a positive definite function* $v(\mathsf{x}, t)$ *whose total derivative* $\dot{v}$ *for* (38.1) *has the following property: the functional* $\dot{v}$ *is nonpositive* (*or* not identically equal to zero) along every solution $x(t)$ for which the condition

$$v(x(s), s) \leq v(x(t), t) \qquad (t_0 \leq s \leq t) \tag{38.15}$$

is satisfied. If v is even decrescent and if $\dot{v}$ is negative definite along the solutions which satisfy (38.15), then the equilibrium is asymptotically stable.

Example: Let

$$\dot{x}(t) = -ax(t) - bx(t - h)$$

If $v = x^2$, the total derivative is

$$\dot{v} = -2x(ax + by) \qquad (y = x(t - h))$$

The condition (38.15) is here of the form

$$x(s)^2 \leq x(t)^2 \qquad (0 \leq s \leq t)$$

It is satisfied by those solutions for which $|x(t - h)| \leq |x(t)|$, i.e., for which $|y| \leq |x|$. In this case it is sufficient to consider the values $s \leq t - h$. Consequently, the inequality $|b| \leq a$, $a > 0$, is a sufficient condition for asymptotic stability of the equilibrium.

Several other conditions can be established instead of (38.15) (Razumichin [4], Krasovskii [18]).

Razumichin [7] showed that criteria for stability according to the first approximation can be derived from Theorem 38.6. Let the system (38.1) be of such a nature that its equation of the first approximation is of the form

$$\dot{x}(t) = A(t) x(t) + B(t) x(t - h) \tag{38.16}$$

with only *one* delay term. Let the ordinary differential equation

$$\dot{x}(t) = (A(t) + B(t)) x(t) \tag{38.17}$$

have an exponentially stable equilibrium, and let

$$v(x, t) = x^T P(t) x$$

be a Liapunov function for (38.17) formed according to Theorem 24.5. Then the derivative for (38.16) is of the form

$$\dot{v}_{(38.16)} = x^T(t) Q(t) x(t) + x^T(t) R(t) x(t - h) \tag{38.18}$$

The condition of Theorem 38.6 requires that (38.18) be negative definite for all those $x(t - h)$ for which the inequality

$$v(x(t - h), t - h) \leq v(x(t), t) \tag{38.19}$$

is satisfied. From this, a sufficient condition for stability according to the first approximation is obtained if the maximum of (38.18) is estimated under the side-condition (38.19) (Razumichin [7]).

Krasovskii [23] pointed out that the stability theory for differential-difference equations can be established in Banach spaces in analogy to the theory in the Euclidean space.

Instead of (38.1), more general differential-difference equations

$$\dot{x}_i = f_i(x_1(t-\vartheta), \ldots, x_n(t-\vartheta); t) \qquad (i = 1, \ldots, n) \qquad (38.20)$$

can be investigated, where the f_i are functionals with respect to the first n arguments, and where ϑ varies between 0 and h (Krasovskii [24]), Hale [1], Šimanov [3, 4]). The core of the method remains unchanged. To investigate the stability according to the stability of the first approximation, the linear part of (38.20) is written by means of Stieltjes integrals in the form

$$\dot{x}_i = \sum_j \int_0^h x_j(t-\vartheta) \, dy_{ij}(\vartheta, t)$$

In the autonomous case, the characteristic equation is given by

$$\det\left(\int_0^h e^{\lambda\vartheta} \, dy_{ij}(\vartheta) - \lambda\delta_{ij}\right) = 0 \qquad (38.21)$$

Šimanov investigated the critical case of a vanishing root of (38.21) by means of the method described in Sec. 30 (a).

Note that in general the considerations of this section cannot be applied without further assumptions to differential-difference equations of *neutral* type. In equations of neutral type (which have received only little attention as yet) derivatives with time-lagged arguments occur in addition to $\dot{x}(t)$, i.e., there are derivatives of the form $\dot{x}(t-h)$, etc. The stability behavior of equations of this type is quite different from that of common differential-difference equations. For example, asymptotic stability for an autonomous system need not be uniform, not even in the linear case. This was shown by Hahn [5] for a linear differential-difference equation of the order n. This equation has a family of solutions of the form

$$e^{\lambda_k(t-t_0)}$$

where

$$\lambda_k = \pm 2\pi i(k+a) - bk^{-\alpha} + o(k^{-\alpha})$$

$$(a > 0, \ b > 0, \ \alpha > 0, \ k = 1, 2, \ldots)$$

The situation is similar to that of example (36.4). The operator mentioned above, which transforms in the case of a linear differential-difference equation z into p, is in the neutral case not necessarily bounded. Consequently, the proof of the equivalence of uniform asymptotic and exponential stability cannot be transferred.

39. APPLICATION OF THE DIRECT METHOD TO DIFFERENCE EQUATIONS

We consider a difference equation of the form

$$\Theta x = f(x, t) \qquad (39.1)$$

SEC. 39 GENERALIZATIONS OF THE CONCEPT OF STABILITY 147

where t is the independent variable, and where the operator $\Theta x(t) \equiv x(t+1)$ denotes the transition from t to $t+1$ in all arguments. Let the function $f(x, t)$ be real and continuous in $|x| < h$ with respect to $|x|$ for fixed t. In addition, let $f(0, t) \equiv 0$. Let the equation and its solution be defined either for a sequence of discrete t-values

$$t_0, t_0 + 1, t_0 + 2, \ldots \qquad (39.2)$$

or for all $t \geq t_0$. In the first case, the solution is determined by the initial point x_0. In the second case, the solution is, in a similar manner as (38.7), determined by an initial function. But, since the value of the solution at an instant t_1 depends only on the value of the initial function at a single point, the second case does not contain any basically new features. Therefore, it is sufficient to restrict the considerations to the first case. The initial instant has to be considered as an integral variable parameter, whose values are contained in a fixed sequence of numbers $t^*, t^* + 1, \ldots$.

The fundamental definitions of Secs. 2 and 17 of the stability of an equilibrium, etc., can readily be transferred to difference equations. However, as in the case of differential-difference equations, a solution defined for the initial instant t_1 cannot always be regarded as a continuation of a solution beginning at $t_0 < t_1$. Consequently, it has to be postulated that the stability definition is independent of the initial instant.

Only a few investigations of the stability behavior of difference equations have been carried out as yet. Without using the direct method, Perron [3] and Ta Li [1] found the first fundamental results. A Liapunov function was first applied in a special investigation by Meredith [1]; however, it cannot be decided whether the direct method was known to the author. Krasovskii [13] transferred some results on differential equations to difference equations. A systematic application of the direct method was presented by Hahn [8].

In order to study the stability behavior of the equilibrium of (39.1), a Liapunov function $v(x, t)$ is used which need be defined with respect to t for the values (39.2) only. Inequalities (1.8) and (1.9), by which the concepts of "definiteness" and "decrescentness" are defined, need as well hold for these t-values only. Instead of the total derivative $\dot{v}$, the "total" difference

$$\Delta v = \Theta v - v = v(\Theta x, t+1) - v(x, t) \qquad (39.3)$$

has to be considered. Disregarding these modifications, Theorems 4.1, 4.2, 5.1, and 5.2 remain valid verbatim. This is also true for the proofs. The integrals occurring in these proofs, of course, have to be replaced by the inverse operation of (39.3), i.e., by the sums. For the autonomous linear difference equation

$$\Theta x = Ax \qquad (39.4)$$

a Liapunov function can easily be constructed similarly as for differential equation (8.1). The Liapunov function is chosen as

$$v(x) = x^T B x \qquad (39.5)$$

The symmetric matrix B is determined by the equation

$$A^T B A - A = -C \qquad (39.6)$$

in analogy to (8.3); C is a positive definite symmetric matrix. Then

$$\Delta v = -x^T C x$$

The quadratic form (39.5) is positive definite if and only if the absolute values of all eigenvalues of A are smaller than 1. Applying this form as in Sec. 8 to the investigation of (39.4), one arrives at Theorems 8.1 to 8.3 with the single difference being that the term "eigenvalue" has to be replaced by "logarithm of the eigenvalue." Therefore, the equilibrium of (39.4) is asymptotically stable if all eigenvalues are smaller than 1 in absolute value, etc. As in Sec. 10, the stability of the perturbed equation can also be studied; in particular, Theorem 10.1 on the stability according to the first approximation remains valid. (This theorem was first proved by Perron [3] by a different method.)

The problem of reversing the stability theorems has not yet been investigated.

If the equilibrium of the nonautonomous linear difference equation

$$\Theta x = A(t) x \qquad (A(t) \text{ bounded and nonsingular}) \qquad (39.7)$$

is exponentially stable, then a Liapunov function can easily be determined by means of the procedure of Theorem 24.5. If $x^T C(t) x$ is a positive definite quadratic form with bounded coefficient matrix, then

$$v(x, t) = \sum_{k=0}^{\infty} p^T(t + k, x, t) C(t + k) p(t + k, x, t) \qquad (39.8)$$

is a positive definite decrescent form with $\Delta v = -x^T C x$ (Hahn [8]). The theorem on stability according to the first approximation (the analogue to Theorem 26.2) can also be proved for nonautonomous difference equations by means of the function (39.8). Ta Li [1] proved this theorem in a different, but equivalent, form (it corresponds to the criterion of Perron [4], mentioned in Sec. 26).

By considering relation (24.2) as a difference equation for the fundamental matrix X, or for the rows of this matrix, it is possible to derive from the theorem on the equilibrium of (39.4) a statement on the stability behavior of the equilibrium of the periodic differential equation (24.1): The stability behavior is determined by the eigenvalues of the matrix S. The eigenvalues must have absolute values smaller than 1 for the equilibrium of (24.2), or (24.1), respectively, to be asymptotically stable. This is one of the statements of Theorem 24.2. The statement on instability is obtained accordingly. By this approach, the main theorem on the stability behavior of linear periodic differential equations can be obtained by means of the direct method.

To find an approximate solution of the differential equation

$$\dot{x} = g(x, t) \tag{39.9}$$

by the so-called "difference method," the difference equation

$$x(t + w) = wg(x, t) + x \tag{39.10}$$

is associated with (39.9). w is the span of the difference process. Here, the following theorem holds.

THEOREM 39.1 (Skalkina [1], Krasovskii [13]): If the equilibrium of differential equation (39.9) is exponentially stable and if the span w is sufficiently small, then the equilibrium of the difference equation (39.10) is also exponentially stable.

Proof: A Liapunov function is chosen for (39.9) which satisfies the estimates of Theorem 22.1. Its total difference

$$\Delta v = v(x + wg, t + w) - v(x, t)$$

for (39.10) can be put into the form

$$\Delta v = \sum_{i=1}^{n} w \left(\frac{\partial v}{\partial x_i}\right)_0 g_i(x, t) + w \left(\frac{\partial v}{\partial t}\right)_0$$

by means of the mean-value theorem. The subscript 0 indicates that the arguments $x + \delta wg$ and $t + \delta w (0 < \delta < 1)$ have to be substituted into the partial derivatives. Because of the assumptions on v, the expressions Δv and

$$\dot{v}_{(39.9)} = \sum_{i=1}^{n} \frac{\partial v}{\partial x_i} g_i(x, t) + \frac{\partial v}{\partial t}$$

have the same sign for sufficiently small w and sufficiently small $|x|$. However, the derivative $\dot{v}$ is negative definite. Consequently, Δv is negative; this means that the equilibrium of (39.10) is asymptotically stable. That it is even exponentially stable can be concluded (by using Theorem 22.1 for difference equations) from the fact that v is also a Liapunov function for (39.10). The existence of such a function guarantees the exponential stability of the equilibrium.

The above mentioned theorems can be applied to the theory of discontinuous control systems which are described by means of difference equations (R. E. Kalman and J. E. Bertram [1]). A theoretical application was considered by Karasik [1] who used Liapunov functions to derive conditions for the existence of periodic solutions of difference equations.

BIBLIOGRAPHY

REMARKS:

1. *DAN:* Doklady Akademii Nauk SSSR, Moscow,
   PMM: Prikladnaja Matematika i Mekhanika, Moscow,
   VMU: Vestnik Moskovskogo Universiteta, Serija Fizikomatematiceskich i estestvennich Nauk.

2. The following periodicals are available in cover-to-cover translations in the English language:
   Astronom Zurn.: Soviet Astronomy (AJ), since January 1957,
   Avtomatika i Telemekhanika: Automation and Remote Control, since January 1956,
   DAN: Soviet Mathematics—Doklady, since January 1960,
   PMM: Applied Mathematics and Mechanics (PMM), since January 1958,
   Usp. Mat. Nauk: Russian Mathematical Surveys, since January 1960.

3. Wherever possible, the English translations of Russian titles and the spelling of the names of Russian authors have been copied from the reference journal "Mathematical Reviews."

4. An asterisk at the name of an author indicates that the corresponding papers or books are published in Russian. Exceptions are indicated by adding in parentheses the language in which the publication is written.

5. Page numbers, etc., of Russian publications refer to the original Russian edition.

*Aizerman, M. A.:
   [1] "On a problem concerning the stability 'in the large' of dynamical systems." *Usp. mat Nauk* **4,** No. 4, 187–188 (1949),
   [2] "On the determination of the safe and unsafe parts on the boundary of stability." *PMM* **14,** 444–448 (1950),
   [3] "Sufficient conditions for the stability of a class of dynamic systems with variable parameters." *PMM* **15,** 382–384 (1951).
   [4] *Theory of Automatic Control of Motors.* Moscow 1952.

*――― and Gantmakher, F. R.:
 [1] "The stability (based on linear approximation) of the periodic solution of the system of differential equations having discontinuous right hand sides." *PMM* **21**, 658–669 (1957), [Announced in *DAN* **116**, 527–530 (1957)],
 [2] "On the stability of the equilibrium position for discontinuous systems." *PMM* **24**, 406–421 (1960).

*Aminov, M. S.:
 [1] "On the stability of certain mechanical systems." *PMM* **12**, 643–646 (1948),
 [2] "On a method for obtaining sufficient conditions for stability of unsteady motions." *PMM* **19**, 621–622 (1955).

Andronow, A. A. and Witt, A.:
 [1] "Zur Stabilität nach Liapounow." *Phys. Z. Sowjetunion* **4**, 606–608 (1933).

*Anosov, D. V.:
 [1] "On stability of equilibrium relay systems." *Avtomatika i Telemekhanika* **20**, 135–149 (1959).

Antosiewicz, H. A.:
 [1] "Stable systems of differential equations with integrable perturbation term." *J. London Math. Soc.* **31**, 208–212 (1956),
 [2] "A survey on Liapunov's second method." *Ann. Math. Studies* **41** (Contr. theory nonlin. oscill. 4), 141–166 (1958).

――― and Davis, P.:
 [1] "Some implications of Liapunov's conditions of stability." *J. Rat. Mech. Analysis* **3**, 447–457 (1954).

*Barbašin, E. A.:
 [1] "The method of sections in the theory of dynamical systems." *Mat. Sbornik* (2) **29**, 233–280 (1951),
 [2] "Stability of the solution of a certain nonlinear third-order equation." *PMM* **16**, 629–632 (1952).

*――― and Krasovskii, N. N.:
 [1] "On stability of motion in the large." *DAN* **86**, 453–456 (1952),
 [2] "On the existence of Liapunov functions in the case of asymptotic stability in the large." *PMM* **18**, 345–350 (1954).

*――― and Skalkina, M. A.:
 [1] "On stability in the first approximation." *PMM* **19**, 623–624 (1955).

Bass, R. W.:
 [1] "Comment on the paper of A. M. Letov." *Regelungstechnik, Moderne Theorien und ihre Verwendbarkeit*, Verlag R. Oldenbourg, München 1957, 209–210.

*Bautin, N. N.:
 [1] "On the behavior of dynamical systems with small violations of the condition of stability of Routh-Hurwitz." *PMM* **12**, 613–632 (1948),

[2] "Criteria for unsafe and safe bounds of a region of stability." *PMM* **12**, 691–728 (1948).

*Bedel'baev, A. K.:
  [1] "On a construction of the functions of Liapunov as a quadratic form." *Izv. Akad. Nauk Kazach. SSR.*, Ser. Mat. Mekh. No. 4 (8), 27–37 (1956),
  [2] "Stability of nonlinear systems of automatic control." Alama-Ata, *Izv. Akad. Nauk Kazach. SSR 1960*, 163 p.

*Beleckii, V. V.:
  [1] "Some problems of the motion of a rigid body in a Newtonian force field." *PMM* **21**, 749–758 (1957).

Bellman, R.:
  [1] "Kronecker products and the second method of Liapunov." *Math. Nachr.* **20,** 17–19 (1959).

Bendixson, H.:
  [1] "Sur les courbes définies par des équations différentielles." *Acta math.* **24**, 1–88 (1900).

*Berezkin, E. N.:
  [1] "Some questions of stability of motion." *VMU* (Ser. Mat.) **11**, No. 1, 23–31 (1956),
  [2] "On stability of unperturbed motion of a mechanical system." *PMM* **23**, 864–871 (1959).

*Blichevskii, V. C.:
  [1] "Conditions for the absence of uniform and asymptotic stability." *Usp. Mat. Nauk* **14,** 1 (85), 141–146 (1959).

*Bromberg, P. V.:
  [1] "On the problem of stability of a class of nonlinear systems." *PMM* **14**, 561–562 (1950).

Cartwright, M. L.:
  [1] "The stability of solution of certain equations of the fourth order." *Quart. J. Mech. Appl. Math.* **9**, 185–194 (1956).

Cesari, L.:
  [1] *Asymptotic Behavior and Stability Problems in Ordinary Differential Equations.* Springer Verlag, Heidelberg 1959 (Ergeb. Math. 16).

Chalanaj, cf. Halanaj.

*Chang, Su-Ying:
  [1] "A stability criterion for non-linear control systems." *Avtomatika i Telemekhanika* **20**, 669–672 (1959).

*Charasachal, V.:
  [1] "On stability of solutions of countably many systems of differential equations by the first approximation." *Izv. Akad. Nauk Kazach. SSR.*, Ser. Mat. Mekh., **60**, No. 3, 77–84 (1949),

[2] "On stability of systems of linear differential equations of second order." *Trudy Sect. Mat. Mekh. Akad. Nauk Kazach. SSR.* **1,** 46–49 (1958).

*Charlamov, P. V.:
  [1] "A case of integrability of the equations of motion of a rigid body in a liquid." *PMM* **19,** 231–233 (1955).

*Chetaev, N. G.:
  [1] "Un théorèm sur l'instabilité." *C. R. (Doklady) Acad. Sci. URSS* **1934 I,** 529–531 (1934) (French),
  [2] "On instability of the equilibrium in certain cases where the force function does not have a maximum." *Uch. Zapiski Kazansk. Univ.* 1938,
  [3] "The smallest characteristic number." *PMM* **9,** 193–196 (1945) (English summary),
  [4] "On the sign of the smallest characteristic number." *PMM* **12,** 101–102 (1948),
  [5] "On the choice of parameters of a stable mechanical system." *PMM* **15,** 371–372 (1951),
  [6] "On unstable equilibrium in certain cases when the force function is not maximum." *PMM* **16,** 89–93 (1952),
  [7] "On stability of rotation of a rigid body with a fixed point in Lagrange's case." *PMM* **18,** 123–124 (1954),
  [8] *The Stability of Motion.* Pergamon Press, Ltd., London 1961 (English),
  [9] "On a gyroscope mounted in a universal suspension (on gimbals)." *PMM* **22,** 379–381 (1958),
  [10] "On certain questions relative to the problem of the stability of unsteady motion." *PMM* **24,** 6–9 (1960),
  [11] "On the stability of rough systems." *PMM* **24,** 20–22 (1960).

Corduneanu, C.:
  [1] "Sur la stabilité asymptotique." *An. Şti. Univ. Iaşi*, Sect. I. **5,** 37–39 (1959),
  [2] "Sur la stabilité asymptotique, II." *Rev. math. pur. appl.* **5,** 573–576 (1960),
  [3] "Application des inégalités différentielles à la théorie de la stabilité." *Abn. Şti. Univ. Iaşi*, Sect. I, **6,** 46–58 (1960) (Russian, French summary).

Cunningham, W. J.:
  [1] *Introduction to Nonlinear Analysis.* McGraw-Hill, New York 1958.

*Czan, Sy-Ni:
  [1] "On stability of motion for a finite interval of time." *PMM* **23,** 230–238 (1959),
  [2] "On stability of motion of a gyroscope." *PMM* **23,** 604–605 (1959),
  [3] "On estimates of solutions of systems of differential equations, accumulation of perturbation and stability of motion during a finite time interval." *PMM* **23,** 640–649 (1959),
  [4] "On the theory of quality of nonlinear control systems." *PMM* **23,** 971–974 (1959).

*Dubošin, G. N.:
  [1] "Essay on the investigation of stability of solutions of nonholomorph systems." *Mat. Sbornik* **42,** 601–612 (1935) (English summary),

[2] "On the stability of solutions of canonical systems." *C. R. (Doklady) Acad Sci. URSS* **1935 I**, 276–287 (1935) (English),

[3] "Sur certaines conditions de la stabilité pour l'équation $\ddot{x} + px = 0$." *C. R. (Doklady) Acad. Sci. URSS* **1935 I**, 390–392 (1935) (French),

[4] "On the problem of stability of a motion under constantly acting perturbations." *Trudy gos. astron. Inst.* Sternberg **14**, No. 1 (1940),

[5] "Some remarks on the theorems of Liapunov's second method." *VMU* **5**, No. 10, 27–31 (1950),

[6] "A stability problem for constantly acting disturbances." *VMU* **7**, No. 2, 35–40 (1952),

[7] *Foundations of the Theory of Stability of Motions*, Moscow 1957.

*Duvakin, A. P. and Letov, A. M.:
[1] "On the stability of regulating systems with two organs of regulation." *PMM* **18**, 162–166 (1954).

*Elsgolts, A. E.:
[1] "Stability of solutions of differential-difference equations." *Usp. Mat. Nauk* **9**, No. 4, 95–112 (1954),

[2] *Qualitative Methods in Mathematical Analysis*, Moscow 1955.

Ergen, W. K., Lipkin, H. J. and Nohel, J. A.:
[1] "Applications of Liapunov's second method in reactor dynamics." *J. Math. Phys.* **36**, 36–48 (1957).

*Eršov, B. A.:
[1] "On stability in the large of a certain system of automatic regulation." *PMM* **17**, 61–72 (1953),

[2] "A theorem on stability of motion in the large." *PMM* **18**, 381–383 (1954).

*Erugin, N. P.:
[1] "On asymptotic stability of the solutions of a certain system of differential equations." *PMM* **12**, 157–164 (1948),

[2] "On certain questions of stability of motion and the qualitative theory of differential equations." *PMM* **14**, 459–512 (1950),

[3] "A qualitative investigation of integral curves of a system of differential equations." *PMM* **14**, 659–664 (1950),

[4] "Theorems on instability." *PMM* **16**, 355–361 (1952),

[5] "On a problem of stability of systems of automatic regulation." *PMM* **16**, 620–628 (1952),

[6] "The methods of A. M. Liapunov and questions of stability in the large." *PMM* **17**, 389–400 (1953),

[7] "Qualitative methods in theory of stability." *PMM* **19**, 599–616 (1955),

[8] "Methods for solving questions of stability in the large." *Transactions of the second All-union Congress on the Theory of Automatic Control*, Moscow **1955, I**, 133–141 (1955).

Ezeilo, J. O. C.:
[1] "On the stability of solutions of certain differential equations of the third order." *Quart. J. Math.*, Oxford Ser. (2), **11**, 64–69 (1960).

*Feldbaum, A. A.:
  [1] "Integral criteria for quality of a regulation." *Avtomatika i Telemekhanika* **9**, 3–19 (1948).

*Filippov, A. F.:
  [1] "Differential equations with discontinuous right hand side." *Mat. Sbornik* **51**, 99–128 (1960).

*Germaidze, V. E.:
  [1] "On asymptotic stability by the first approximation." *PMM* **21**, 133–135 (1957),
  [2] "On asymptotic stability of systems with lagging argument." *Usp. Mat. Nauk* **14**, No. 4 (88), 149–156 (1959).

*—— and Krasovskii, N. N.:
  [1] "On stability under constant perturbations." *PMM* **21**, 769–774 (1957).

*Gorbunov, A. D.:
  [1] "On conditions for monotonic stability of a system of ordinary linear homogeneous differential equations." *VMU* **6**, No. 3, 15–24 (1951),
  [2] "On certain properties of the solutions of a system of ordinary linear homogeneous differential equations." *VMU* **6**, No. 6, 3–16 (1951),
  [3] "On conditions for asymptotic stability of the zero-solution of a system of ordinary linear homogeneous differential equations." *VMU* **8**, No. 9, 46–55 (1953),
  [4] "Estimates for the characteristic exponent of the solutions of a system of ordinary linear homogeneous differential equations." *VMU* **11**, No. 2, 7–13 (1956).

*Goršin, S.:
  [1] "On the stability of motion under constantly acting perturbations." *Izv. Akad. Nauk Kazach. SSR.* **56**, Ser. Mat. Mekh., No. 2, 46–73 (1948),
  [2] "On stability of solutions of a system of countably many differential equations under constantly acting perturbations." *Izv. Akad. Nauk Kazach. SSR* **60**, Ser. Mat. Mekh., No. 3, 32–38 (1949),
  [3] "On Liapunov's second method." *Izv. Akad. Nauk Kazach. SSR.* **97**, No. 4 (1950).

*Gradštein, I. S.:
  [1] "Application of A. M. Liapunov's theory of stability to the theory of differential equations with small coefficients in the derivatives." *Mat. Sbornik* (2) **32**, 262–286 (1952) [announced in *Usp. Mat. Nauk* **6**, No. 6, 156–157 (1951)].

Hahn, W.:
  [1] "Bericht über Differential-Differenzengleichungen mit festen und veränderlichen Spannen." *J. Ber. deutsche Math.-Verein* **57**, 55–84 (1954),
  [2] "Über Stabilität bei nichtlinearen Systemen." *Z. angew. Math. Mech.* **35**, 459–462 (1955),
  [3] "Eine Bemerkung zur zweiten Methode von Ljapunov." *Math. Nachr.* **14**, 349–354 (1956),

[4] "Behandlung von Stabilitätsproblemen mit der zweiten Methode von Ljapunov." *Beiheft "Nichtlineare Regelungsvorgänge" der "Regelungstechnik,"* 51–66 (1956),
[5] "Über Differential-Differenzengleichungen mit anomalen Lösungen." *Math. Ann.* **133**, 251–255 (1957),
[6] "Probleme und Methoden der modernen Stabilitätstheorie." *MTW-Mitt. TH Wien* **4**, 119–134 (1957),
[7] "Bemerkungen zu einer Arbeit von Herrn Vejvoda." *Math. Nachr.* **20**, 21–24 (1959),
[8] "Über die Anwendung der Methode von Ljapunov auf Differenzengleichungen." *Math. Ann.* **136**, 430–441 (1958).

Halanaj (Chalanaj), A.:
[1] "Stabilitätssätze für Differentialgleichungen mit verzögertem Argument." *Rev. math. pur. appl.* **3**, 207–215 (1958),
[2] "Periodische und fastperiodische Lösungen von Differentialgleichungssystemen mit verzögertem Argument." *Rev. math. pur. appl.* **4**, 685–691 (1959),
[3] "Stabilitätskriterien für Differentialgleichungssysteme mit verzögertem Argument." *Rev. math. pur. appl.* **5**, 367–374 (1960),
[4] "Integrale Stabilität für Differentialgleichungssysteme mit verzögertem Argument." *Rev. math. pur. appl.* **5**, 541–548 (1960).

Hale, J. K.:
[1] "Asymptotic behavior of the solutions of differential difference equations." *RIAS techn. Report* 61-10, 33 p. (1961).

Herschel, R.:
[1] "Über ein verallgemeinertes quadratisches Optimum." *Regelungstechnik* **4**, 190–195 (1956),
[2] "Über die Grenzen des quadratischen Optimums." *Regelungstechnik* **5**, 469–472 (1957).

Ingwerson, D. R.:
[1] "A modified Liapunov method for nonlinear stability analysis." *IRE Trans. Automatic Control* **6**, 199–210 (1961); Discussion, *IRE Trans. Automatic Control* **7**, 85–88 (1962).

*Isaeva, I. C.:
[1] "On the sufficient conditions of rotational stability of the 'tip-top' gyroscope placed on an absolutely rough horizontal plane." *PMM* **23**, 403–406 (1959).

*Kalinin, S. V.:
[1] "On the stability of periodic motions in the case when one root is equal to zero." *PMM* **12**, 671–672 (1948),
[2] "On the stability of periodic motions in the case when the characteristic equation has a pair of purely imaginary roots (abridged procedure)." *PMM* **21**, 125–128 (1957),
[3] "On stability of motion of an airplane with autopilot." *Izv. Akad. Nauk SSSR. otd. techn. Nauk* **1958**, No. 4, 114–117.

Kalman, R. E. and Bertram, J. E.:
  [1] "Control analysis and design via the 'second method' of Liapunov. I. Continuous-time systems. II. Discrete-time systems." *ASME J. of Basic Engineering*, 371–393, 394–400 (1960).

*Kamenkov, G. V.:
  [1] "Sur la stabilité du mouvement dans un cas particulier." *Sbornik nauč. trud. av. inst. Kazan* **4,** 3–18 (1935) (French),
  [2] "On stability of motion." *Sbornik nauč. trud. av. Inst. Kazan* **9** (1939),
  [3] "On stability of motion over a finite interval of time." *PMM* **17,** 529–540 (1953).

*―――― and Lebedev, A. A.:
  [1] "Remark to a paper on stability over a finite interval of time." *PMM* **18,** 512 (1954).

Kamke, E.:
  [1] *Differentialgleichungen reeller Funktionen*, 2. Ed., Leipzig 1948.

*Karasik, G. Ia.:
  [1] "On conditions for the existence of periodic solutions of difference equations." *Izv. vyss. ucebn. zaved.* **4,** No. 11, 70–79 (1959).

*Kartvelishvili, N. A.:
  [1] "Stability in the large in a water power plant with storage basin under stationary conditions." *Inž. Sbornik* **20,** 25–30 (1954).

*Kats, I. Ia. and Krasovskii, N. N.:
  [1] "On stability of a system with random parameters." *PMM* **24,** 809–823 (1960).

*Kim, Če Džen:
  [1] "Determination of the region of influence of an asymptotically stable position of equilibrium." *VMU* **14,** No. 2, 3–14 (1959).

Kolmogorov, A. N. and Fomin, S. V.:
  [1] *Elements of the Theory of Functions and Functional Analysis;* Vol. 1, *Metric and normed spaces.* Graylock Press, Rochester, N. Y., 1957.

*Komarnitskaia, O. I.:
  [1] "Stability of nonlinear automatic control systems." *PMM* **23,** 505–514 (1959).

*Krasovskii, N. N.:
  [1] "Theorems on stability of motions determined by a system of two equations." *PMM* **16,** 547–554 (1952),
  [2] "On stability of the solutions of a nonlinear system of three equations for arbitrary initial disturbances." *PMM* **17,** 339–350 (1953),
  [3] "On stability of solutions of a system of two differential equations." *PMM* **17,** 651–672 (1953),
  [4] "On a problem of stability of motion in the large." *DAN* **88,** 401–404 (1953),

[5] "Stability of solutions of second-order systems in critical cases." *DAN* **93**, 965–967 (1953),

[6] "On stability of motion in the large for constantly acting disturbances." *PMM* **18**, 95–102 (1954),

[7] "On the behavior in the large of integral curves of a system of two differential equations." *PMM* **18**, 149–154 (1954),

[8] "On the inversion of theorems of A. M. Liapunov and N. G. Chetaev on instability for stationary systems of differential equations." *PMM* **18**, 513–532 (1954),

[9] "On stability in the large of the solutions of a nonlinear system of differential equations." *PMM* **18**, 735–737 (1954),

[10] "Sufficient conditions for stability of solutions of a system of nonlinear differential equations." *DAN* **98**, 901–904 (1954),

[11] "On stability of motion in the critical case of a single zero root." *Mat. Sbornik* (2) **37**, 83–88 (1955),

[12] "On inversion of K. P. Persidskii's theorem on uniform stability." *PMM* **19**, 273–278 (1955),

[13] "On stability in the first approximation." *PMM* **19**, 516–530 (1955),

[14] "On conditions of inversion of A. M. Liapunov's theorems on instability for stationary systems of differential equations." *DAN* **101**, 17–20 (1955),

[15] "On the theory of Liapunov's second method in studying the steadiness of motion." *DAN* **109**, 460–463 (1956) (preliminary note, cf. [16]),

[16] "Inversion of theorems on Liapunov's second method and questions of stability of motion in the first approximation." *PMM* **20**, 255–265 (1956),

[17] "Concerning the application of A. M. Liapunov's method for equations with lags." *PMM* **20**, 315–327 (1956),

[18] "On asymptotic stability of systems with after effect." *PMM* **20**, 513–518 (1956),

[19] "On the inversion of theorems of the second method of A. M. Liapunov for investigation of stability of motion." *Usp. Mat. Nauk* **9**, No. 3, 159–164 (1956),

[20] "On the theory of the second method of A. M. Liapunov for the investigation of stability." *Mat. Sbornik* **40** (82), 57–64 (1956),

[21] "On stability with large initial perturbations." *PMM* **21**, 309–319 (1957),

[22] "On periodic solutions of differential equations involving a time lag." *DAN* **114**, 252–255 (1957),

[23] "The stability of quasi-linear systems with after effects." *DAN* **119**, 435–438 (1958),

[24] *Certain Problems of the Theory of Stability of Motion.* Gos. Izd. Fiz.-mat. Lit. Moscow 1959, 211 p.,

[25] "On the theory of optimum control." *PMM* **23**, 624–639 (1959),

[26] "On optimum control in the presence of random disturbances." *PMM* **24**, 64–79 (1960).

*Krementulo, V. V.:

[1] "Investigation of the stability of a gyroscope, taking account of dry friction on the axis of the inner gimbal ring." *PMM* **23**, 968–970 (1959),

[2] "Stability of a gyroscope having a vertical axis of the outer ring with dry friction in the gimbal axes taken into account." *PMM* **24,** 568–571 (1960).

*Kudakova, P. V.:
[1] "On stability in a finite time interval." *Akad. Nauk Kazach. SSR. Trudy, Sect. mat. mekh.* **1,** 41–45 (1958).

*Kunin, I. A.:
[1] "Determination of a finite region of initial deviations for which the motions remain asymptotically stable for a system of two equations of the first order." *PMM* **16,** 539–546 (1952).

*Kurzweil, J.:
[1] "On the reversibility of the first theorem of Liapunov concerning the stability of motion." *Czechoslov. math. J.* **5** (80), 382–398 (1955) (English summary),
[2] "On the reversibility of the second theorem of Liapunov concerning the stability of motion." *Czechoslov. math. J.* **5** (80), 435–438 (1955) (English summary, preliminary note, cf. [3]),
[3] "The converse second Liapunov's theorem concerning the stability of motion." *Czechoslov. math. J.* **6** (81), 217–259, 455–473 (1956) (English summary).

*——— and Vrkoč, I.:
[1] "The converse theorems of Liapunov and Persidskii concerning the stability of motion." *Czechoslov. math. J.* **7** (82), 254–274 (1957) (English summary).

*Kuškov, N. N.:
[1] "Theorems on limit cycles for a system of ordinary differential equations." *Usp. Mat. Nauk* **13, 2** (80), 203–209 (1958).

*Kuzmin, P. A.:
[1] "On the theory of stability of motion." *PMM* **18,** 125–127 (1954).

LaSalle, J. P.:
[1] "The extent of asymptotic stability." *Proc. mat. Acad. Sci. USA* **48,** 363–365 (1960),
[2] "Some extensions of Liapunov's second method." *IRE Trans. profess. group circuit theory* CT-7, 520–527 (1960).

——— and Lefschetz, S.:
[1] *Stability by Liapunov's Direct Method with Applications.* Academic Press, New York 1961.

*Lebedev, A. A.:
[1] "The problem of stability in a finite interval of time." *PMM* **18,** 75–94 (1954),
[2] "On stability of motion during a given interval of time." *PMM* **18,** 139–148 (1954),
[3] "On a method of constructing Liapunov functions." *PMM* **21,** 121–124 (1957),
[4] "Stability of motion in a finite interval of time." *Moscov. Ord. Lenina Aviac. Inst. Trudy* **112,** 106–113 (1959).

Lefschetz, S.:
- [1] *Differential Equations, Geometric Theory.* Interscience, New York 1957,
- [2] "Liapunov and stability of motion." *Bol. Soc. mat. Mexicana* (2) **3**, 25–39 (1958),
- [3] "Controls: An application of the direct method of Liapunov." *Symposium intern. ecuac. dif. ord. Mexico*, 139–143 (1961),
- [4] "On automatic controls." *RIAS techn. Rep.* 60–9, 6 p. (1960).

Lehnigk, S. H.:
- [1] "Über quadratische Formen mit Parametern als Ljapunovsche Trägerfunktionen." *DFL-Bericht* **106**, 36 p. (1958),
- [2] "On Liapunov's second method with parameter-dependent quadratic forms in the case of autonomous nonlinear equations which have a linear part." *Proc. First Intern. Congress of the Intern. Federation of Automatic Control*, 934–938 (1960).

*Letov, A. M.:
- [1] "On the theory of an isodromic regulator." *PMM* **12**, 363–368 (1948),
- [2] "The regulation of a stationary state of a system subjected to constant perturbing forces." *PMM* **12**, 149–156 (1948),
- [3] "On a special case in the investigation of the stability of a system of regulation." *PMM* **12**, 729–736 (1948),
- [4] "Naturally unstable systems under control." *PMM* **14**, 183–192 (1950),
- [5] "Bounds for the smallest characteristic value of a class of regulating systems." *PMM* **15**, 591–600 (1951),
- [6] "Stability of control systems with two regulating organs." *PMM* **17**, 401–410 (1953),
- [7] "Stability of unsteady motions of control systems." *PMM* **19**, 257–264 (1955),
- [8] *Stability in Nonlinear Control Systems.* Princeton Univ. Press, Princeton, N. J., 1961 (English),
- [9] "The status of the problem of stability in the theory of automatic control (a survey)." *Trudy II, vsesojuz. sov. teor. avtom. reg.*, **1**, 79–104 (1955),
- [10] "Die Stabilität von Regelsystemen mit nachgebender Rückführung." *Regelungstechnik, Moderne Theorien und ihre Verwendbarkeit*, Verlag R. Oldenbourg, München 1957, 201–210 (Russian and German, English summary),
- [11] "Stability and quality of nonlinear systems of automatic control" (Chap. 1 of *Problems of the Theory of Nonlinear Systems of Automatic Regulation and Control*, edited by Ia. Z. Cypkin, Moscow 1957),
- [12] "Stability of an automatically controlled bicycle, rolling on a horizontal plane." *PMM* **23**, 650–655 (1959).

    Remark: [8] contains the results of most of the earlier publications of the author.

Levinson, N.:
- [1] "On a nonlinear differential equation of the second order." *J. math. Phys. Massachusetts* **22**, 181–187 (1943).

*Liapunov, M. A.:
  [1] "Problème général de la stabilité du mouvement." *Ann. Fac. Sci. Toulouse* **9**, 203–474 (1907) (French, translation of the original paper published in 1893 in *Comm. Soc. math. Kharkow*; reprinted as Vol. 17 in *Ann. math. Studies*, Princeton 1949),
  [2] "Investigation of a singular case of the problem of stability of motion." *Mat. Sbornik* **17**, 252–333 (1893).

*Liaščhenko, N. Ia.:
  [1] "On asymptotic stability of solutions of a system of differential equations." *DAN* **96**, 237–239 (1954).

*Litvartovskii, I. V.:
  [1] "Certain questions of stability in the first approximation for differential equations with discontinuous right hand sides." *Moscovsk. fiz.-techn. Inst. Isledov. Mekh. priklad. Mat. Trudy* **3**, 47–63 (1959),
  [2] "Some tests for the stability of the solutions of a system of differential equations with discontinuous right hand members." *DAN* **125**, 733–736 (1959),
  [3] "On stability of the solutions of a system of differential equations with discontinuous right hand sides." *PMM* **23**, 598–603 (1959).

*Lur'e, A. I.:
  [1] "On stability of one type of systems under control." *PMM* **9**, 353–367 (1945) (English summary),
  [2] "Investigation of the stability of motion of a dynamic system." *PMM* **11**, 445–448 (1947) (English summary),
  [3] "On the canonical form of the equations of the theory of automatic regulation." *PMM* **12**, 651–666 (1948),
  [4] "On the character of the bounds of the region of stability of regulating systems." *PMM* **14**, 371–382 (1950),
  [5] "On the problem of stability of regulating systems." *PMM* **15**, 67–74 (1951),
  [6] "On strictly unstable regulating systems." *PMM* **15**, 251–254 (1951),
  [7] *Einige nichtlineare Probleme aus der Theorie der automatischen Regelung.* Akademie-Verlag, Berlin 1957 (German),
  [8] "Liapunov's direct method and its application to the theory of automatic control." *Trudy II, vsesojuz. sov. teor. avtom. reg.*, **1**, 142–148 (1955).

*———— and Postnikov, V. N.:
  [1] "Concerning the stability of regulating systems." *PMM* **8**, 246–248 (1944).

Magnus, K.:
  [1] "Das A-Kurvenverfahren zur Berechnung nichtlinearer Regelungsvorgänge." *Beiheft "Nichtlineare Regelungsvorgänge" der "Regelungstechnik,"* 9–28 (1956),
  [2] "On the stability of a heavy symmetrical gyroscope on gimbals." *PMM* **22**, 173–178 (1958) (Russian).

*Maizel, A. D.:
  [1] "On stability in the first approximation." *PMM* **14**, 171–182 (1950),

[2] "On stability of solutions of systems of differential equations." *Ural. polytechn. Inst. Trudy* **51**, 20–60 (1946).

*Makarov, I. P.:
 [1] "Quelques généralizations des théorèmes fondamentaux de Liapounoff sur la stabilité du mouvement." *Bull. Soc. phys.-math. Kazan III* **10**, 139–159 (1938) (French summary).

*Malkin, I. G.:
 [1] "Das Existenzproblem von Ljapunovschen Funktionen." *Izv. fiz.-mat. Obsc. Kazan III* **4**, 51–62 (German) and III **5**, 63–84 (1931),
 [2] "Über die Stabilität der Bewegung in Sinne von Liapounoff." *C. R. (Doklady) Acad. Sci. URSS* (2) **15**, 437–439 (1937) (German, preliminary note, cf. [6]),
 [3] *Certain Questions in the Theory of Stability of Motion in the Sense of Liapunov.* American Math. Soc., No. 20, New York 1950 (English),
 [4] "Über die Bewegungsstabilität nach der ersten Näherung." *C. R. (Doklady) Acad. Sci. URSS* (2) **18**, 159–162 (1938) (German),
 [5] "Verallgemeinerung des Fundamentalsatzes von Liapounoff über die Stabilität der Bewegungen." *C. R. (Doklady) Acad. Sci. URSS* (2) **18**, 162–164 (1938) (German),
 [6] "On the stability of motion in the sense of Liapunov." *Mat. Sbornik* **3**, 47–100 (1938) (German summary),
 [7] "Sur un théorème d'existence de Poincaré-Liapounoff." *C. R. (Doklady) Acad. Sci. URSS* (2) **27**, 307–310 (1940) (French),
 [8] "Basic theorems of the theory of stability of motion." *PMM* **6**, 411–448 (1942) (English summary),
 [9] "Stability in the case of constantly acting disturbances." *PMM* **8**, 241–245 (1944) (English summary),
 [10] "On the theory of stability of regulating systems." *PMM* **15**, 59–66 (1951),
 [11] "On the solution of a stability problem in the case of two purely imaginary roots." *PMM* **15**, 255–257 (1951),
 [12] "On a method of solution of the problem of stability in the critical case of a pair of purely imaginary roots." *PMM* **15**, 473–484 (1951),
 [13] "Solution of some critical cases of the problem of stability of motion." *PMM* **15**, 575–590 (1951),
 [14] "A theorem on stability in the first approximation." *DAN* **76**, 783–784 (1952),
 [15] "On the construction of Liapunov functions for systems of linear equations." *PMM* **16**, 239–242 (1952),
 [16] "On a problem of the theory of stability of systems of automatic regulation." *PMM* **16**, 365–368 (1952),
 [17] "On the stability of systems of automatic regulation." *PMM* **16**, 495–499 (1952),
 [18] "On a theorem concerning stability of motion." *DAN* **84**, 877–878 (1952).
 [19] *Theorie der Stabilität einer Bewegung.* Verlag R. Oldenbourg, München 1959 (German) (A very poor English translation of this book is *"Theory of*

Stability of Motion," Atomic Energy Commission, Translation No. 3352, Dept. of Commerce, Washington, D. C. 1958),

[20] "On the reversibility of Liapunov's theorem on asymptotic stability." *PMM* **18**, 129–138 (1954).

Remark: [19] contains many of the earlier results found by the author.

*Marachkov, V.:
[1] "On a theorem of Liapunov." *Bull. Soc. phys.-math. Kazan III*, **12**, 171–174 (1940) (German summary).

Massera, J. L.:
[1] "On Liapounoff's condition of stability." *Ann. of Math.* (2) **50**, 705–721 (1949),
[2] "Total stability and approximately periodic vibrations." *Fac. Ing. Montevideo, Publ. Inst. Mat. Estad.* **2**, 135–145 (1954) (Spanish, English summary),
[3] "Sobre la estabilidad en espacios de dimension infinta." *Rev: Un. mat. Argentina* **17**, 135–147 (1955),
[4] "Contributions to stability theory." *Ann. of Math.* (2) **64**, 182–206 (1956). Correction, *Ann. of Math.* (2) **68**, 202 (1958),
[5] "On the existence of Liapunov functions." *Publ. Inst. Mat. Estad. Montevideo* **3**, No. 4, 111–124 (1960),
[6] "Converse theorems of Liapunov's second method." *Symposium intern. ecuac. dif. ord. Mexico*, 158–163 (1961).

—— and Schäffer, J. J.:
[1] "Linear differential equations and functional analysis I." *Ann. of Math.* (2) **67**, 517–573 (1958),
[2] ". . . II." *Ann. of Math.* (2) **69**, 88–104 (1959),
[3] ". . . III." *Ann. of Math.* (2) **69**, 555–574 (1959),
[4] ". . . IV." *Math. Ann.* **139**, 287–342 (1960).

*Melnikov, G. I.:
[1] "Some problems concerning Liapunov's direct method." *DAN* **110**, 326–329 (1956).

Meredith, C. A.:
[1] "On a nonlinear difference equation." *J. London Math. Soc.* **15**, 260–272 (1942).

Moisseev, N. D.:
[1] "Über den unwesentlichen Charakter einer der Beschränkungen, welche den topographischen Systemen in der Liapounoffschen Stabilitätstheorie auferlegt werden." *C. R. (Doklady) Acad. Sci. URSS* 1936 **I**, 165–166,
[2] "Über die Wahrscheinlichkeit der Stabilität nach Liapounoff." *C. R. (Doklady) Acad. Sci. URSS* 1936 **I**, 215–217,
[3] "Über die Stabilität der Lösungen eines Systems von Differentialgleichungen." *Math. Ann.* **113**, 452–460 (1936),
[4] "Über Stabilitätswahrscheinlichkeitsrechnung." *Math. Z.* **42**, 513–537 (1937),
[5] *Summary of the History of Stability.* Moscow 1949 (Russian).

*Morosova, E. P.:
  [1] "Stability of rotation of a solid suspended on a string." *PMM* **20,** 621-626 (1956).

*Movchan, A. A.:
  [1] "The direct method of Liapunov in stability problems of elastic systems." *PMM* **23,** 483-493 (1959),
  [2] "Stability of processes with respect to two metrics." *PMM* **24,** 988-1001 (1960).

*Mufti, I. H.:
  [1] "Stability in the large of systems of two equations." *Arch. rat. Mech. Anal.* **7,** 118-134 (1961).

*Myškis, A. D.:
  [1] "General theory of differential equations with lagged argument." *Usp. Mat. Nauk* **4,** No. 5, 99-141 (1949).

*Nemyckii, V. V.:
  [1] "Some problems of the qualitative theory of differential equations (Review of modern literature)." *Usp. Mat. Nauk* **9,** No. 3, 39-56 (1954),
  [2] "Estimate of the regions of asymptotic stability of nonlinear systems." *DAN* **101,** 803-804 (1954),
  [3] "Liapunov's method of rotating functions for finding oscillatory regimes." *DAN* **97,** 33-36 (1954),
  [4] "On certain methods for the qualitative investigation of many dimensional autonomous systems in the large." *Trudy Moscovsk. mat. Obsc* **5,** 455-482 (1956).

——— (Nemitzky, V. V.) and Stepanov, V. V.:
  [1] *Qualitative Theory of Differential Equations.* Princeton Univ. Press, Princeton, N. J., 1960.

Nougmanova, Ch.:
  [1] "Sur la stabilité des mouvements périodiques." *C. R. (Doklady) Acad. Sci. URSS* **42,** 202-204 (1944).

*Ogurstov, A. I.:
  [1] "The stability of two differential equations of the third and fourth order." *PMM* **23,** 179-181 (1959).

Perron, O.:
  [1] "Über Stabilität und asymptotisches Verhalten der Integrale von Differentialgleichungssystemen." *Math. Z.* **29,** 129-160 (1928),
  [2] "Die Ordnungszahlen linearer Differentialgleichungssysteme." *Math. Z.* **31,** 748-766 (1929),
  [3] "Über Stabilität und asymptotisches Verhalten der Lösungen eines Systems endlicher Differenzengleichungen." *J. reine angew. Math.* **161,** 41-61 (1929),
  [4] "Die Stabilitätsfrage bei Differentialgleichungen." *Math. Z.* **32,** 703-728 (1930).

*Persidskii, K. P.:
  [1] "Au sujet du problème de stabilité." *Bull. Soc. phys.-math. Kazan III* **5**, No. 3, 56–62 (1931) (French),
  [2] "On stability in the first approximation." *Mat. Sbornik* **40**, 284–293 (1933) (German summary),
  [3] "Un theorème sur la stabilité du mouvement." *Bull. Soc. phys.-math. Kazan III* **6**, 76–79 (1934) (French),
  [4] "On the stability theory of the solutions of systems of differential equations." *Bull. Soc. phys.-math. Kazan III* **8**, (1936),
  [5] "On a theorem of Liapunov." *C. R. (Doklady) Acad. Sci. URSS* **14**, 541–544 (1937),
  [6] "On the theory of stability of solutions of differential equations." Thesis, Moscow 1946, summary: *Usp. mat. Nauk* **1**, No. 1, 5–6, 250–255 (1946),
  [7] "On the stability of the solutions of an infinite system of equations." *PMM* **12**, 597–612 (1948),
  [8] "On stability of solutions of a system of countably many differential equations." *Izv. Akad. Nauk Kazach. SSR* **56**, Ser. Mat. Mekh., No. 2, 3–35 (1948),
  [9] "Countable systems of differential equations and the stability of their solutions." *Uč. Zapiski Kazach. Gos. Univ. Mat. Fiz:* No. 2 (1949),
  [10] "Uniform stability in the first approximation." *PMM* **13**, 229–240 (1949),
  [11] "On stability of solutions of differential equations." *Izv. Akad. Nauk Kazach. SSR* **60**, Ser. Mat. Mekh., No. 4, 3–18 (1950),
  [12] "On Liapunov's second method in linear normed spaces." *Vestnik Akad. Nauk Kazach. SSR* **1958**, No. 7, 89–97,
  [13] "Inversion of Liapunov's second theorem on instability in linear normed spaces." *Vestnik Akad. Nauk Kazach. SSR* **1959**, No. 10, 31–35.

*Persidskii, S. K.:
  [1] "On the second method of Liapunov." *Izv. Akad. Nauk Kazach. SSR* **1956**, No. 4 (8), 43–47,
  [2] "On stability in a finite interval." *Vestnik Akad. Nauk Kazach. SSR* **1959**, No. 9, 75–80,
  [3] "Some theorems on the second method of Liapunov." *Vestnik Akad. Nauk Kazach. SSR* **1960**, No. 2, 70–76,
  [4] "On Liapunov's second method." *PMM* **25**, 17–23 (1961).

Pestel, E.:
  [1] "Anwendung der Ljapunovschen Methode und des Verfahrens von Krylov-Bogoljubov auf ein technisches Beispiel." *Beiheft "Nichtlineare Regelungsvorgänge" der "Regelungstechnik,"* 67–85 (1956).

*Pirogov, I. Z.:
  [1] "On the stability of a gyroscopic system." *PMM* **23**, 1134–1136 (1959).

*Pliškin, Iu. M.:
  [1] "On the question of the estimate of the integral criteria of the quality of the regulation of a nonlinear system." *Avtomatika i Telemekhanika* **16**, 19–26 (1955).

*Pliss, V. A.:
 [1] "A qualitative picture of the integral curves in the large and the construction with arbitrary accuracy of the region of stability of a certain system of two differential equations." *PMM* **17,** 541–554 (1954),
 [2] "An investigation of a nonlinear system of three differential equations." *DAN* **117,** 184–187 (1957),
 [3] "The necessary and sufficient conditions for the stability in the whole of a homogeneous system of three differential equations." *DAN* **120,** 708–710 (1958),
 [4] "Aizerman's problem in the case of three simultaneous differential equations." *DAN* **121,** 422–425 (1958),
 [5] *Certain Problems of the Theory of Stability of Motion in the Whole.* Izd. Leningradsk. Univ. 1958, 182 p.

Popov, E. P.:
 [1] *Dynamik der Systeme automatischer Regelung.* Berlin 1957.

*Popov, V. M.:
 [1] "On relaxation of sufficient conditions of absolute stability." *Avtomatika i Telemekhanika* **19,** 3–9 (1958) (English summary).

*Pozarickii, G. K.:
 [1] "On non-steady motion of conservative holonomic systems." *PMM* **20,** 429–433 (1956),
 [2] "On the stability of dissipative systems." *PMM* **21,** 503–512 (1957),
 [3] "On the construction of Liapunov functions from the integrals of the equations of the perturbed motion." *PMM* **22,** 145–154 (1958).

*Razumchin, B. S.:
 [1] "On stability of the trivial solution of systems of ordinary differential equations of second order." *PMM* **19,** 279–288 (1955),
 [2] "On stability of automatic control systems possessing one control unit." *Avtomatika i Telemekhanika* **17,** 958–968 (1956) (English summary),
 [3] "On stability of unsteady motion." *PMM* **20,** 266–270 (1956),
 [4] "On stability of systems with retardation." *PMM* **20,** 500–512 (1956),
 [5] "Estimates of solutions of a system of differential equations of the perturbed motion with variable coefficients." *PMM* **21,** 119–120 (1957),
 [6] "On the equilibrium of systems with a small multiplier." *PMM* **21,** 578–580 (1957),
 [7] "Stability according to the first approximation of systems with a retardation." *PMM* **22,** 155–166 (1958),
 [8] "On the application of Liapunov's method to stability problems." *PMM* **22,** 338–349 (1958).

Reghiş, M.:
 [1] "Sur la stabilité d'après la première approximation." *Lucranile şti. Inst. Ped. Timişoara. Mat.-Fiz.*, 135–143 (1958).

Reissig, G.:
 [1] "Über eine nichtlineare Differentialgleichung zweiter Ordnung." *Math. Nachr.* **13,** 313–318 (1955),

[2] "Über die totale Stabilität erzwungener Reibungsschwingungen." *Abh. Deutsche Akad. Wiss. Kl. Math. Phys. Techn.* 1959, No. 1, 6–28 (1959),
[3] "Einige topologische Fragen im Zusammenhang mit erzwungenen Schwingungen." *M. Ber. Deutsche Akad. Wiss.* **1,** 145–154 (1959),
[4] "Kriterien für die Zugehörigkeit dynamischer Systeme zur Klasse D." *Math. Nachr.* **20,** 67–72 (1959),
[5] "Über die Stabilität linearer Differentialgleichungssysteme zweiter Ordnung." *M. Ber. Deutsche Akad. Wiss.* **2,** 146–148 (1960),
[6] "Über die totale Stabilität erzwungener Bewegungen mit kombinierter Dämpfung." *Z. angew. Math. Mech.* **39,** 401–402 (1959),
[7] "Stabilitätskriterien für ein lineares System mit veränderlichen Parametern." *Z. angew. Math. Mech.* **40,** T112–T113 (1960),
[8] "Stabilitätsprobleme in der qualitativen Theorie der Differentialgleichungen." *J. Ber. Deutsche Math.-Verein.* **63,** 97–116 (1960),
[9] "Neue Probleme und Methoden aus der qualitativen Theorie der Differentialgleichungen." *M. Ber. Deutsche Akad. Wiss.* **2,** 1–8 (1960),
[10] "Ein Kriterium für asymptotische Stabilität." *M. Ber. Deutsche Akad. Wiss.* **2,** 583–587 (1960).

Rekasius, Z. V. and Gibson, J. E.:
[1] "Stability analysis of nonlinear control systems by the second method of Liapunov." *IRE Trans. automatic control* **7,** 3–23 (1962).

Reuter, G. E. H.:
[1] "A boundedness theorem for nonlinear differential equations of the second order." *Proc. Cambridge phil. Soc.* **47,** 49–54 (1951),
[2] "Boundedness theorems for nonlinear differential equations of the second order II." *J. London math. Soc.* **27,** 48–58 (1952).

*Riabov, G. A.:
[1] "On stability of a particular solution of the three-body problem." *Astronom. Žurn.* **29,** 341–349 (1952).

*Roitenberg, J. N.:
[1] "On a method of constructing Liapunov functions for linear systems with variable coefficients." *PMM* **22,** 167–172 (1958).

*Rozenvasser, E. N.:
[1] "Stability of nonlinear control systems described by differential equations of the 5th and 6th order." *Avtomatika i Telemekhanika* **19,** 101–113 (1958) (English summary).

*Rumiancev, V. V.:
[1] "On stability under the conditions of S. A. Chaplygin of the screw motion of a rigid body in a fluid." *PMM* **19,** 229–230 (1955),
[2] "On stability of permanent rotations of a heavy rigid body." *PMM* **20,** 51–66 (1956),
[3] "On the theory of stability of regulated systems." *PMM* **20,** 714–722 (1956),
[4] "On stability of permanent rotations of a solid body around a fixed point." *PMM* **21,** 339–346 (1957),

[5] "Stability of the rotation of a rigid body having an ellipsoidal cavity filled with fluid." *PMM* **21,** 740–748 (1957),

[6] "On stability of motion with respect to a subset of the variables." *VMU* Ser. Mat. mekh. **1957,** No. 4, 9–16 (1957),

[7] "On the stability of motion of a gyroscope on gimbals." *PMM* **22,** 374–378 (1958),

[8] "On the stability of motion of a gyroscope on gimbals II." *PMM* **22,** 499–503 (1958),

[9] "On the stability of rotational motion of a rigid body with a liquid inclusion." *PMM* **23,** 1057–1065 (1959),

[10] "A theorem on stability of motion." *PMM* **24,** 47–54 (1960),

[11] "On the stability of rotation of a top with a cavity filled with a viscous liquid." *PMM* **24,** 603–609 (1960),

[12] "On stability of motion of a gyroscope." *PMM* **25,** 9–16 (1961).

Saltykow, N. N.:
[1] "Le theorème de Liapounov sur la stabilité des solutions des équations différentielles." *J. Math. pur. appl. IX* **36,** 229–234 (1957).

Sansone, G. and Conti, R.:
[1] *Equazioni Differenziali Non Lineari.* Roma 1956 (Chap. IX).

Schultz, D. G. and Gibson, J. E.:
[1] "The variable gradient method for generating Liapunov functions." To appear in *IRE Trans. antom. control.*

Seibert, P.:
[1] "Ultimate boundedness and stability under perturbations." *RIAS techn. Rep.* 60-7 (1960), 15 p.

*Šestakov, A. A.:
[1] "Some theorems on stability in Liapunov's sense." *DAN* **79,** 25–28 (1951).

*Šimanov, S. N.:
[1] "On stability of solution of a nonlinear equation of the third order." *PMM* **17,** 369–372 (1953),

[2] "On stability of the solutions of a nonlinear system of equations." *Usp. Mat. Nauk* **8,** No. 6, 155–157 (1955),

[3] "On the instability of the motion of a system with retardation." *PMM* **24,** 55–63 (1960),

[4] "On the stability in the critical case of zero root for systems with time lag." *PMM* **24,** 447–457 (1960).

*Skackov, B. N.:
[1] "On the stability of a class of nonlinear systems of automatic regulation." *Vestnik Leningradsk. Univ.* **13,** No. 1, 46–56 (1957) (English summary),

[2] "Questions of stability in the large and regulating properties for a certain system of differential equations." *Vestnik Leningradsk. Univ., Ser. Mat. Mekh. Astron.* **13,** No. 3, 67–80 (1957) (English summary),

[3] "On the region of stability of some nonlinear systems of regulation." *Vestnik Leningradsk. Univ.* **15,** No. 1, 100–103 (1960) (English summary),

[4] "On the stability of some nonlinear systems of differential equations." *Vestnik Leningradsk. Univ.* **15,** No. 7, 164–167 (1960) (English summary).

*Skalkina, M. A.:
[1] "On the preservation of asymptotic stability in transition from differential equations to the corresponding difference equations." *DAN* **104,** 505–508 (1955).

*Skimel', V. V.:
[1] "On problems of stability of motion of a heavy rigid body about a fixed point." *PMM* **20,** 130–132 (1956).

*Spasskii, R. A.:
[1] "On a class of regulated systems." *PMM* **18,** 329–344 (1954).

*Starzinskii, V. M.:
[1] "Sufficient conditions for stability of a mechanical system with one degree of freedom." *PMM* **16,** 369–374 (1952),
[2] "On the stability of unsteady motion in one case." *PMM* **16,** 500–504 (1952),
[3] "On stability of unsteady motions in a special case." *PMM* **19,** 471–480 (1955).

*Štelik, V. G.:
[1] "On the determination of a finite time interval of the stability of solutions of a system of differential equations." *Ukrain. mat. Žurn.* **10,** 100–102 (1958) (English summary),
[2] "On the solutions of a linear system of differential equations with almost periodic coefficients." *Ukrain. mat. Žurn.* **10,** 318–327 (1958) (English summary).

Stepanoff, V.:
[1] "Zur Definition der Stabilitäts-Wahrscheinlichkeit." *C. R. (Doklady) Acad. Sci. URSS* **18,** 151–154 (1938).

*Tabarovskii, A. M.:
[1] "On the stability of motion of a heavy gyroscope on gimbals." *PMM* **24,** 572–574 (1960),
[2] "On the stability of motion of Foucault gyroscopes with two degrees of freedom." *PMM* **24,** 796–801 (1960).

Ta Li:
[1] "Die Stabilitätsfrage bei Differenzengleichungen." *Acta math.* **63,** 99–141 (1934).

Taussky, O.:
[1] "A remark on a theorem of Liapunov." *J. math. Analysis and Appl.* **2,** 105–107 (1961).

*Troickii, V. A.:
[1] "On the canonical transformations of the equations of the theory of automatic regulation." *PMM* **17,** 49–60 (1953),

[2] "On the behavior of dynamical systems and systems of automatic regulation having several regulating organs near to the boundary of a region of stability." *PMM* **17,** 673–684 (1953).

*Tuzov, A. P.:
[1] "Stability questions for a certain regulating system." *Vestnik Leningradsk. Univ., Ser. mat.-fiz.* **10,** No. 22, 43–70 (1955),
[2] "On the stability in the large of a certain regulation system." *Vestnik Leningradsk. Univ., Ser. mat.-fiz.* **12,** No. 1, 57–75 (1957) (English summary).

Vejvoda, O.:
[1] "The stability of a system of differential equations in the complex domain." *Czechoslov. math. J.* **7** (82), 137–159 (1957) (Czech, English and Russian summaries).

Veksler, D.:
[1] "Stability theorems for a system of stationary differential equations." *Rev. Math. pur. appl.* **3,** 131–138 (1958).

*Volkov, D. M.:
[1] "An analogue of the second method of Liapunov for nonlinear boundary value problems of hyperbolic equations." *Leningradsk. gos. Univ. Uc. Zapiski 271*, Ser. mat. Nauk **33,** 90–96 (1958).

Vorel, Z.:
[1] "On some applications of Liapounoff's theory in electrical machinery." *Aplicare Matematiky, Prague* 1956, 59–78 (Czech, English and Russian summaries).

*Vorovich, I. I.:
[1] "On the stability of motion with random disturbances." *Izv. Akad. Nauk SSSR, Ser. mat.* **20,** 17–32 (1956).

Vrkoč, J.:
[1] "On the inverse theorem of Chetaev." *Czechoslov. math. J.* **5** (80), 451–461 (1955) (Czech, English summary),
[2] "Integral stability." *Czechoslov. math. Zurn.* **9** (84), 71–129 (1959) (Russian, English summary).

Wright, E. M.:
[1] "The linear difference differential equation with constant coefficients." *Proc. roy. Soc. Edinburgh* **A62,** 387–393 (1949).

*Yakubovich (Jakubovič), V. A.:
[1] "A criterion for the reducibility of a system of differential equations." *DAN* **66,** 577–580 (1948),
[2] "On a class of nonlinear differential equations." *DAN* **117,** 44–46 (1957),
[3] "On the boundedness and stability in the whole of the solutions of some nonlinear differential equations." *DAN* **121,** 984–986 (1958),
[4] "Stability condition in the large for some nonlinear differential equations of automatic control." *DAN* **135,** 26–29 (1960).

Yoshizava, T.:
- [1] "On the stability of solutions of a system of differential equations." *Mem. Coll. Sci. Univ. Kyoto A* **29**, 27–33 (1955),
- [2] "Note on the boundedness of solutions of a system of differential equations." *Mem. Coll. Sci. Kyoto A* **28**, 293–298 (1954),
- [3] "Note on the solutions of a system of differential equations." *Mem. Coll. Sci. Kyoto A* **29**, 249–273 (1955),
- [4] "Note on the boundedness and the equiultimate boundedness of solutions of $x' = F(t, x)$." *Mem. Coll. Sci. Kyoto A* **29**, 275–291 (1955), **30**, 91–103 (1957),
- [5] "On the necessary and sufficient conditions for the uniform boundedness of solutions of $x' = F(t, x)$." *Mem. Coll. Sci. Kyoto A* **30**, 217–226 (1957),
- [6] "Note on the equiultimate boundedness of solutions of $x' = F(t, x)$." *Mem. Coll. Sci. Kyoto A* **31**, 211–217 (1958),
- [7] "On the equiasymptotic stability in the large." *Mem. Coll. Sci. A* **32**, 171–180 (1959),
- [8] "Liapunov's functions and boundedness of solutions." *RIAS techn. Rep.* 59–7 (1959), 10 p.,
- [9] "Liapunov's functions and boundedness of solutions." *Funkcialaj Ekvacioj* **2**, 95–142 (1959),
- [10] "Existence of a bounded solution and existence of a periodic solution of the differential equation of the second order." *RIAS techn. Rep.* 60–18 (1960), 13 p.,
- [11] "Stability and boundedness of systems." *Arch. rat. Mech. Anal.* **6**, 409–421 (1960).

*Zhak (Žak), S. V.:
- [1] "On the stability of certain unique cases of the movement of a symmetric gyroscopic, containing liquid masses." *PMM* **22**, 245–249 (1958).

*Zheleznov (Železnov), E. I.:
- [1] "Sufficient conditions for limit cycles." *Izv. vyšs. učebn. Zaved., Mat.* **1**, 127–132 (1957).

*Zubov (Zoobow), V. I.:
- [1] "Some sufficient criteria for stability of a nonlinear system of differential equations." *PMM* **17**, 506–508 (1953),
- [2] "On the theory of A. M. Liapunov's second method." *DAN* **99**, 341–344 (1954),
- [3] "Questions of the theory of Liapunov's second method, construction of a general solution in the region of asymptotic stability." *PMM* **19**, 179–210 (1955),
- [4] "On the theory of A. M. Liapunov's second method." *DAN* **100**, 857–859 (1955),
- [5] "An investigation of the stability problem of systems of equations with homogeneous right hand members." *DAN* **114**, 942–944 (1957),
- [6] *The Methods of Liapunov and their Applications.* Leningrad 1957,

- [7] "Conditions for asymptotic stability in the case of nonstationary motion and estimate of the rate of decrease of the general solution." *Vestnik Leningradsk. Univ.*, Ser. Mat. Mekh. Astron. **12,** No. 1, 110–129 (1957),
- [8] "On a method of investigating the stability of a null-solution in doubtful cases." *PMM* **22,** 46–49 (1958),
- [9] "On stability conditions in a finite time interval and on the computation of the length of that interval." *Bull. Inst. Politechn. Iasi, N. S.* **4** (8), 69–74 (1958) (German summary),
- [10] *Mathematical Methods for the Investigation of Systems of Automatic Control.* Gos. Sojus. Izd. Sudostroit. Promysl., Leningrad 1959, 324 p.,
- [11] "Some problems in stability of motion." *Mat. Sbornik* **48** (90), 149–190 (1959).

AUTHOR INDEX

A

Aizerman, 8, 34–36, 40, 42–44, 102
Aminov, 26, 122
Andronow, 123
Anosov, 35
Antosiewicz, v, 6, 16, 63, 72, 95, 108

B

Barbašin, 4, 8, 15, 16, 47, 61, 64, 65, 69, 71, 73, 78, 101, 142
Bass, 55
Bautin, 124
Bedel'baev, 29, 57
Beleckii, 25
Bellman, 26
Bendixson, 12
Berezkin, 116, 122
Bertram, v, 150
Blichevskii, 75
Bromberg, 52

C

Cartwright, 48
Cesari, v
Chalanaj (cf. Halanaj)
Chang, 55
Charasachal, 32, 136
Charlamov, 25
Chetaev, 11, 12, 16, 18, 20, 24, 25, 31, 35, 41, 56, 74, 75, 105–107
Conti, vi
Corduneanu, 16, 66, 103, 110
Cunningham, v
Czan, 26, 129

D

Davis, 95
Dubošin, 7, 20, 24, 32, 107, 112, 125
Duvakin, 55

E

Elsgolts, vii, 140–142
Ergen, 22
Eršov, 109
Erugin, 20, 35, 43, 44, 46, 71, 79
Ezeilo, 48

F

Feldbaum, 59
Filippov, 35
Fomin, 133, 135

G

Gantmakher, 35, 102
Germaidze, 103, 109, 142, 143
Gibson, 39, 55
Gorbunov, 57, 95–97, 127
Goršin, 108, 136
Gradštein, 65, 96

H

Hahn, 26–28, 34, 36, 62, 64, 124, 139, 146–148
Halanaj (Chalanaj), 141–143
Hale, 146
Herschel, 59

I

Ingwerson, 38, 39
Isaeva, 26

J

Jakubovič (cf. Yakubovich)

K

Kalinin, 122, 123
Kalman, v, 150

Kamenkov, 120, 121, 127, 128
Kamke, 9, 87, 98
Karasik, 150
Kartvelishvili, 36
Kats, 10, 18
Kim, 39
Kolmogorov, 133, 135
Komarnitskaia, 54
Krasovskii, 4, 8, 10, 15–18, 23, 37–39, 44–47, 58, 61, 63–66, 69, 71–75, 85, 86, 88, 91, 100, 101, 103, 104, 109, 114, 135, 141–147, 149
Krementulo, 26
Kudakova, 129
Kunin, 36
Kurzweil, 9, 61, 65, 67–69, 71
Kuškov, 131
Kuzmin, 16, 95

L

LaSalle, v, 8, 18, 129
Lebedev, 30, 126–129
Lefschetz, v, vi, 18, 49, 55, 78
Lehnigk, 35
Letov, 49, 50, 53–55, 57, 128
Levinson, 12
Liapunov, vi, 1, 11, 14, 15, 19, 24, 29, 32–34, 75, 84, 93, 96–98, 105, 111, 113–115, 121, 122, 126, 129
Liaščenko, 95
Lipkin, 22
Litvartovskii, 35
Lur'e, 49, 50, 52–55, 124, 128

M

Magnus, 25, 125
Maizel, 107
Makarov, 17, 20
Malkin, vi, 16, 29, 35, 42–44, 53, 60, 63, 65, 70, 92, 95, 103, 104, 107, 108, 112, 114–120, 124, 140
Marachkov, 16
Massera, 8, 16, 20, 60, 61, 63, 65–67, 70–73, 85, 103, 104, 106, 109, 110, 135, 136, 142
Melnikov, 23
Meredith, 147
Moisseev, 1, 9, 15, 29, 116
Morosova, 25

Movchan, 136, 138
Mufti, 46
Myškis, 139

N

Nemyckii (Nemitzky), vii, 13, 39, 46
Nemitzky (cf. Nemyckii)
Nohel, 22
Nougmanova, 32

O

Ogurstov, 48

P

Perron, vii, 34, 35, 39, 84, 98, 99, 102, 103, 107, 147, 148
Persidskii, K. P., 60, 61, 65, 69, 75, 85, 94, 96, 103, 136
Persidskii, S. K., 20, 66, 72, 75, 129, 136
Pestel, 34
Pirogov, 26
Pliškin, 58
Pliss, 44, 46
Popov, 53
Postnikov, 49
Pozarickii, 24, 25, 102

R

Razumchin, 30, 31, 41, 48, 55, 57, 96, 144, 145
Reghiş, 103
Reissig, 10, 13, 32, 131
Rekasius, 55
Reuter, 12, 129
Riabov, 41
Roitenberg, 32
Rozenvasser, 53
Rumiancev, 10, 25, 26, 55

S

Saltykow, 34
Sansone, vi
Schäffer, 103, 136

AUTHOR INDEX 177

Schultz, 39
Seibert, 130
Šestakov, 104
Šimanov, 48, 146
Skackov, 43, 56
Skalkina, 101, 149
Skimel', 25
Spasskii, 50, 55
Starzinskii, 39, 40
Štelik, 102, 129
Stepanoff (Stepanov), vii, 10
Stepanov (cf. Stepanoff)

T

Tabarovskii, 26
Ta Li, 147, 148
Taussky, 26
Troickii, 50, 53, 124
Tuzov, 16, 46

V

Vejvoda, 9, 28, 116, 121
Veksler, 17
Volkov, 139

Vorel, 39
Vorovich, 55, 108
Vrkoč, 61, 69, 74, 109

W

Witt, 123
Wright, 144

Y

Yakubovich (Jakubovič), 49, 55, 93
Yoshizava, 10, 67, 69, 129, 131, 132

Z

Žak (cf. Zhak)
Zhak (Žak), 25
Železnov (cf. Zheleznov)
Zheleznov (Železnov), 131
Zoobow (cf. Zubov)
Zubov (Zoobow), vii, 10, 38, 39, 61, 67, 72, 75, 76, 78–81, 91, 92, 109, 112, 120, 133, 134, 137, 138

SUBJECT INDEX

A

Absolutely stable, 49, 52, 61
Adjoint differential equation, 98
Aizerman's problem, 42–46, 55
Almost periodic, 13, 67, 102
Asymptotically stable, 6, 8, 15–17, 21, 26, 37, 38, 61, 76, 77, 95, 103–106, 113, 114, 120, 121, 134, 135, 138–140, 143, 145, 148
  according to the mth approximation, 118
  equi~, 63, 129
  exponentially, 85
  in the large, 8
  in the whole, 8, 15, 37, 39, 81, 135
  quasi , 6–8, 16, 61, 129
    equi~, 63
  uniformly, 61, 63–65, 76, 77, 110, 134, 135, 140, 141, 143
    according to the first approximation, 85
    in the whole, 64, 65
    with respect to the spatial coordinates, 63
Attracting, uniformly, 67, 134
Attraction, domain of, 8, 38, 78–81, 108
Autonomous difference equation, 147
Autonomous differential-difference equation, 146
Autonomous differential equation, 3, 60, 62, 64, 70, 73, 74, 77, 78, 91, 93, 98, 101, 104, 108, 111, 112, 122, 123, 135

B

Boundary, stability, 80, 81, 123, 124
Bounded:
  equi-distance , 132
  ultimately, 129
  uniform-distance , 132
  uniformly, 129, 130
    ultimately, 129, 130

Branch of a motion, 2

C

Canonical form, 50
Canonical system, 24, 101
Characteristic equation, 42, 92, 93, 112, 122, 123, 144
Characteristic number, 29, 84
Class C_0, C_r, $\bar{C}_0$, $\bar{C}_r$, $C\omega$, K, 2
Class E, 3
Comparison function, 62
Comparison system, 42
Complete differential equation, 33, 101, 103, 111, 117–119
Complete general system, 133
Completely exponentially unstable, 85
Completely stable, 8
Completely unstable, 9, 19, 20, 27, 136
Complex equations of motion, 9, 17, 28, 116
Conditions for nonlinearities, 23
Control:
  optimal, 23
  quality, 57, 58
  system, 43, 49, 58, 150
Converse theorems, 60, 68
Critical behavior, 27, 123
Critical case, 27, 111, 146
Critical eigenvalue, 112, 121
Cycle, 12

D

Dangerous domain, 124
Decrescent, 4, 5, 141, 147
Decrescentness (cf. Decrescent)
Definite, 3, 4, 141, 147
  first integral, 24
  negative, 3, 4
  positive, 3, 4
Delay term, 139

SUBJECT INDEX 179

Derivative, total, 5, 147
  for an equation, 5
Diameter of a domain, 124
Difference equation, 93, 146–148
  autonomous, 147
  linear, 147
  nonautonomous, 148
Difference, total, 147
  for an equation, 147
Differential-difference equation, 139, 146, 147
  autonomous, 146
  neutral, 146
Differential equation:
  adjoint, 98
  autonomous, 3, 60, 62, 64, 70, 73, 74, 77, 78, 91, 93, 98, 101, 104, 108, 111, 112, 122, 123, 135
  complete, 33, 101, 103, 111, 117–119
  in the variations (cf. Equation)
  linear, 3, 26, 29, 33, 92–99, 101, 103, 110
  linearized, 101, 111, 122
  linearly perturbed, 103
  nonautonomous, 12, 74, 77, 78, 81, 92, 93
  of the first approximation (cf. Equation)
  of the perturbed motion, 7, 123
  periodic, 3, 60, 62, 64, 70, 92, 93, 101, 104, 109, 122, 123, 148, 149
  perturbed, 103, 104, 107
  reduced, 101, 112, 117–120
  reducible, 98
  regular, 98, 104, 107
  unperturbed, 103
  with slowly varying coefficients, 30
Direct method of Liapunov, 11
Discontinuous Liapunov function, 13
Distance, 9, 12, 76, 133
Distance-bounded:
  equi-, 132
  uniform-, 132
Domain, 2
  dangerous, 124
  diameter of a, 124
  instability, 74
  nondangerous, 124
  of attraction, 8, 38, 78–81, 108
  stability, 81, 124
    of the initial points, 22, 123
    of the parameters, 22, 123
  $v < 0$, 18
Dynamical system, 61, 71, 75, 76, 78, 137

E

Eigenvalue:
  critical, 112, 121
  noncritical, 112
Equation:
  characteristic, 42, 92, 93, 112, 122, 123, 144
  in the variations, 102
  of the first approximation, 33, 101, 102, 106, 123, 128
Equiasymptotically stable, 63, 129
  quasi-, 63
Equi-distance-bounded, 132
Equilibrium, 1, 3, 7
  isolated, 3, 7
  stability, 1
Equistable, 67, 132
Estimate, 22, 23
Exponentially asymptotically stable, 85
Exponentially stable, 17, 68, 85, 92, 94, 95, 110, 140, 143, 144, 149
Exponentially unstable, 85, 92
  completely, 85

F

First approximation:
  equation of the, 33, 101, 102, 106, 123, 128
  stable according to the, 33, 102, 104, 107, 111, 136, 138, 145, 146, 148
  uniformly, 85
First integral, 24, 25
  definite, 24
First method of Liapunov, 11
Function, Liapunov, 3, 11, 13–15
  discontinuous, 13
  rotating, 13
Functional, Liapunov, 134
Fundamental system, normal, 97

G

General system, 133, 137, 140
  complete, 133
  incomplete, 133
Gyro, 25, 26

180 SUBJECT INDEX

H

Half trajectory, 2
Hurwitz inequalities, 42

I

Incomplete general system, 133
Inherently unstable, 54
Initial function, 139
Initial instant, 3
Initial point, 3, 139
  stability domain, 22, 123
Initial values, 1, 3
Instability, 1, 6, 19 (cf. Unstable)
  domain of, 74
Integrally stable, 109, 142
Intensive behavior, 85, 86, 101, 107
Invariant set, 18, 75, 76, 78, 133, 134
Isolated equilibrium, 7

L

Lagrange:
  coordinates, 1
  stable in the sense of, 129
  theorem of, 24
Liapunov:
  function, 3, 11, 13–15
    discontinuous, 13
    rotating, 13
  functional, 134
Limit equation, 35
Linear difference equation, 147
Linear differential equation, 3, 26, 29, 33, 92–99, 101, 103, 110
Linearization, 33
Linearized differential equation, 101, 111, 122
Linearly perturbed differential equation, 103
Lipschitz condition, 2
Lur'e's problem, 49, 56, 128

M

Mathieu equation, 40

Monotonically stable, 126
Motion, 1, 2, 9, 16, 133
  branch of a, 2
  complex equations of, 9, 17, 28, 116
  perturbed, 5
    differential equation of the, 7, 123
  space, 2
  stability, 1
  unperturbed, 5
mth approximation:
  stable according to the, 117
    asymptotically, 118
  unstable according to the, 118

N

Negative definite, 3, 4
  semi-definite, 3
Neighborhood of a set, 76
Neutral differential-difference equation, 146
Nonautonomous difference equation, 148
Nonautonomous differential equation, 12, 74, 77, 78, 81, 92, 93
Noncritical eigenvalue, 112
Nondangerous domain, 124
Nonlinearities, conditions for, 23
Nonuniformly stable, 61
Normal fundamental system, 97

O

Optimal control, 23
Order number, 57, 84, 96–98, 106
Order numbers of a differential equation, 98

P

Parameter·
  space, 5
  stability domain, 22, 123
Periodic differential equation, 3, 60, 62, 64, 70, 92, 93, 101, 104, 109, 122, 123, 148, 149
Perturbation, stable under constantly acting, 107

SUBJECT INDEX 181

Perturbed differential equation, 103, 104, 107
  linearly, 103
Perturbed motion, 5
  differential equation of the, 7, 123
Phase:
  curve, 2
  space, 2, 4
Positive definite, 3, 4
Positive semi-definite, 3
Probability:
  of stability, 9, 10
  stable with, 10, 18
Problem:
  of Aizerman, 42–46, 55
  of Lur'e, 49, 56, 128

Q

Qualitative method, 44
Quality of control, 57, 58
Quasi-asymptotically stable, 6–8, 16, 61, 129
Quasi-equiasymptotically stable, 63

R

Radially unbounded, 4
Reduced differential equation, 101, 112, 117–120
Reducible, 93
Reducible differential equation, 98
Reduction, 33
Regular differential equation, 98, 104, 107
Rotating Liapunov function, 13

S

Second method of Liapunov, 11
Semi-definite, 3
  negative, 3
  positive, 3
Sensitivity of stability behavior, 100
Significant behavior, 27, 85, 92, 100, 101
Singular case, 114
Singular point, 3
Slowly varying coefficients, equation with, 30

Stability, 1, 6, 7 (cf. Stable)
  boundary, 80, 81, 123, 124
  domain, 81, 124
    of the initial points, 22, 123
    of the parameter, 22, 123
  of an equilibrium, 1
  of motion, 1
  probability of, 9, 10
Stable, 1, 6–8, 14, 16, 17, 24, 38, 61, 76, 133–136, 139, 140, 144, 147
  absolutely, 49, 52, 61
  according to the first approximation, 33, 102, 104, 107, 111, 136, 138, 145, 146, 148
  according to the mth approximation, 117
  asymptotically, 6, 8, 15–17, 21, 26, 37, 38, 61, 76, 77, 95, 103–106, 113, 114, 120, 121, 134, 135, 138–140, 143, 145, 148
    according to the mth approximation, 118
    equi~, 63, 129
    exponentially, 85
    in the large, 8
    in the whole, 8, 15, 37, 39, 81, 135
    quasi-, 6–8, 16, 61, 129
      equi~, 63
    uniformly, 61, 63–65, 76, 77, 110, 134, 135, 140, 141, 143
      according to the first approximation, 85
      in the whole, 64, 65
      with respect to the spatial coordinates, 63
  completely, 8
  equi~, 67, 132
  exponentially, 17, 68, 85, 92, 94, 95, 110, 140, 143, 144, 149
  in a finite interval, 126
  integrally, 109, 142
  in the sense of Lagrange, 129
  monotonically, 126
  nonuniformly, 61
  strongly, 67
  totally, 104, 107, 108, 131, 142
    in the whole, 109
  under constantly acting perturbations, 107
  uniformly, 8, 60–62, 65, 140
    in the whole, 62

Stable (cont.):
   weakly, 6, 28, 60, 68, 114, 122
   with probability, 10, 18
   with respect to a subset of the variables, 10
Strongly stable, 67

T

Three-body problem, 41
Total derivative, 5, 147
   for an equation, 5
Total difference, 147
   for an equation, 147
Totally stable, 104, 107, 108, 131, 142
   in the whole, 109
Trajectory, 1, 2, 75
   half, 2
Trivial solution, 7

U

Ultimately bounded, 129
   uniformly, 129, 130
Unbounded, radially, 4
Uniform-distance-bounded, 132

Uniformly asymptotically stable, 61, 63–65, 76, 77, 110, 134, 135, 140, 141, 143
   according to the first approximation, 85
   in the whole, 64, 65
   with respect to the spatial coordinates, 63
Uniformly attracting, 67, 134
Uniformly bounded, 129, 130
Uniformly stable, 8, 60–62, 65, 140
   in the whole, 62
Uniformly ultimately bounded, 129, 130
Unperturbed differential equation, 103
Unperturbed motion, 5
Unstable, 6, 8, 19, 20, 24, 27, 28, 76, 77, 104–106, 113, 114, 121, 122, 134, 135
   according to the mth approximation, 118
   completely, 9, 19, 20, 27, 136
      exponentially, 85
   exponentially, 85, 92
   in a finite interval, 126
   inherently, 54

W

Weakly intensive behavior, 89
Weakly stable, 6, 28, 60, 68, 114, 122